Table of Contents

Introduction	1
OpenStack and Virtualised HPC	5
OpenStack and HPC Network Fabrics	15
OpenStack and High-Performance Data	27
OpenStack and HPC Workload Management	41
OpenStack and HPC Infrastructure Management	57
OpenStack and Research Cloud Federation	73
Summary	83

AUTHORS

"OpenStack and Virtualised HPC", "OpenStack and HPC Network Fabrics", "OpenStack and High-Performance Data", "OpenStack and HPC Workload Management" and "OpenStack and HPC Infrastructure Management" by Stig Telfer, CTO, StackHPC Ltd, with the support of Cambridge University.

"OpenStack and Research Cloud Federation" by Enol Fernández, Cloud Technologist, EGI.

CONTRIBUTIONS, GUIDANCE AND FEEDBACK

OPENSTACK AND VIRTUALISED HPC:
Blair Bethwaite, *Senior HPC Consultant,* Monash University

Dr. Xiaoyi Lu, *Senior Research Associate,* NOWLAB, The Ohio State University

Dhabaleswar K. (DK) Panda, *Professor and University Distinguished Scholar,* NOWLAB, The Ohio State University

OPENSTACK AND HPC NETWORK FABRICS:
Paul Browne, *OpenStack Engineer,* Cambridge University

Dr. Xiaoyi Lu, *Senior Research Associate,* NOWLAB, The Ohio State University

Jonathan Mills, *Senior HPC Engineer,* CSC @ NASA Goddard Space Flight Center

Dhabaleswar K. (DK) Panda, *Professor and University Distinguished Scholar,* NOWLAB, The Ohio State University

OPENSTACK AND HIGH-PERFORMANCE DATA:
Dr. Tom Connor, *Senior Lecturer,* Cardiff University and the CLIMB collaboration

Vincent Ferretti, *Senior Principal Investigator & Associate Director, Bioinformatics Software Development,* Ontario Institute for Cancer Research

Francois Gerthoffert, *Senior Project Manager,* Ontario Institute for Cancer Research

George Mihaiescu, *Senior Cloud Architect,* Ontario Institute for Cancer Research

Bob Tiernay, *Senior Software Architect,* Ontario Institute for Cancer Research

Andy Yang, *Senior Java Developer,* Ontario Institute for Cancer Research

Christina Yung, *Project Manager, Cancer Genome Collaboratory,* Ontario Institute for Cancer Research

Junjun Zhang, *Project Leader,* Ontario Institute for Cancer Research

The contributors to the "OpenStack and High-Performance Data" section, Ontario Institute for Cancer Research (OICR) case study, wish to acknowledge the funding support from the Discovery Frontiers: Advancing Big Data Science in Genomics Research program (grant number RGPGR/448167-2013, 'The Cancer Genome Collaboratory'), which is jointly funded by the Natural Sciences and Engineering Research Council (NSERC) of Canada, the Canadian Institutes of Health Research (CIHR), Genome Canada, and the Canada Foundation for Innovation (CFI), and with in-kind support from the Ontario Research Fund of the Ministry of Research, Innovation and Science.

James Beal, *Systems Administrator*, **Sanger Institute**
Peter Clapham, *Informatics Support (HPC), Group Leader*, **Sanger Institute**
Dave Holland, *Principal UNIX Systems Administrator*, **Sanger Institute**
Jonathan Nicholson, *HPC Group*, **Sanger Institute**

The Sanger Institute team wish to acknowledge support from Sébastien Buisson, Richard Mansfield and James Coomer from DDN.

OPENSTACK AND HPC WORKLOAD MANAGEMENT:
Lev Lafayette, *HPC Support and Training Officer*, **University of Melbourne**
Dr. Xiaoyi Lu, *Senior Research Associate*, **NOWLAB, The Ohio State University**
Bernard Meade, *Visualization Officer*, **University of Melbourne**
Dhabaleswar K. (DK) Panda, *Professor and University Distinguished Scholar*, **NOWLAB, The Ohio State University**
David Perry, *Computer Integration Specialist*, **University of Melbourne**
Timothy Randles, *Computer Engineer*, **Los Alamos National Laboratory**
Greg Sauter, *HPC on the Cloud and Software Defined Storage Project Manager*, **University of Melbourne**
Daniel Tosello, *Computer Scientist and Research Data Analyst*, **University of Melbourne**
Piotr Wachowicz, *Cloud Integration Team Lead*, **Bright Computing**

OPENSTACK AND HPC INFRASTRUCTURE MANAGEMENT:
Robert Budden, *Senior Cluster Systems Developer*, **Pittsburgh Supercomputer Center**
Jeremy Coles, *Senior Research Associate & SKA/SDP Project Manager*, **Cambridge University**
Kate Keahey, *Scientist*, **Argonne National Laboratory, University of Chicago and Chameleon Science Director**
Pierre Riteau, *Programmer*, **University of Chicago and Chameleon DevOps lead**

OPENSTACK AND RESEARCH CLOUD FEDERATION:
Tim Bell, *Compute and Monitoring Group Leader*, **CERN**
Enol Fernández, *Cloud Technologist*, **EGI**
Kalle Happonen, *Senior Systems Specialist*, **CSC - IT Center for Science Ltd.**
Dan Still, *Development Manager*, **CSC - IT Center for Science Ltd.**
Salman Toor, *Senior Cloud Architect*, **SNIC**
Khalil Yazdi, *Founder*, **Open Research Cloud**

Introduction

OpenStack® is the leading open source IaaS platform, powering many of the world's most notable science and research organisations. Surprisingly, research and science disciplines comprise some of the most prevalent use cases for OpenStack clouds, and OpenStack has provided compelling solutions for many of the challenges of delivering flexible infrastructure for research computing.

High-performance computing (HPC) and high-throughput computing (HTC) workloads require massive scaling and cluster networking; storage, compute and networking access to large volumes of data; and workload and infrastructure manageability. OpenStack software supports these needs today and the development community is rapidly expanding services to fill gaps. By managing resources as an OpenStack private cloud, researchers are able to work in environments tailored to their requirements. The dynamic, automated nature of software-defined infrastructure cuts away time wasted on the distractions of setup, and enables researchers to maximise the time they spend on research itself.

This paper is intended for HPC system architects and research computing managers that are exploring the benefits of cloud and how to bring those benefits to HPC workloads. It can also be used by current OpenStack users to delve into additional capabilities. This is a deep dive into the important functions, considerations, and further reading. Each section includes user examples that describe real-world architecture and operations.

- **OpenStack and HPC Virtualisation.** Describes and addresses the overhead associated with virtualisation.

  > CASE STUDY: Monash University Monash Advanced Research Computing Hybrid (MonARCH) cluster and bursting to the federated NeCTAR Research Cloud.

- **OpenStack and HPC Network Fabrics.** Describes several solutions for delivering unique demands for HPC networking on an OpenStack cloud.

  > CASE STUDY: Cambridge University RDMA (Remote Direct Memory Access)-Centric Bioinformatics Cloud

  > CASE STUDY: The Ohio State University Network-Based Computing Laboratory (NOWLAB)

- **OpenStack and High-Performance Data.** Describes HPC data requirements, OpenStack integration with HPC storage infrastructure, and provides methods to achieve high-performance data management.

    CASE STUDY: Ontario Institute for Cancer Research (OICR) Cancer Genome Collaboratory (The Collaboratory)

    CASE STUDY: The CLoud Infrastructure for Microbial Bioinformatics (CLIMB)

    CASE STUDY: The Wellcome Trust Sanger Institute

- **OpenStack and HPC Workload Management.** Differences between HPC and business cloud workload management, and the benefits OpenStack brings to traditional HPC workload management.

    CASE STUDY: Bright Computing Cluster on Demand

    CASE STUDY: The Ohio State University Network-Based Computing Laboratory (NOWLAB)

    CASE STUDY: Los Alamos National Laboratory (LANL)

    CASE STUDY: University of Melbourne Spartan cluster

- **OpenStack and HPC Infrastructure Management.** Differences, advantages and limitations of managing HPC infrastructure as a bare metal OpenStack cloud.

    CASE STUDY: Chameleon, an experimental testbed for Computer Science, led by University of Chicago, with Texas Advanced Computing Center (TACC), University of Texas at San Antonio (UTSA), Northwestern University and The Ohio State University.

    CASE STUDY: Pittsburgh Supercomputer Center Bridges project, "bridges the research community with HPC and Big Data."

    CASE STUDY: Square Kilometre Array

- **OpenStack and Research Cloud Federation.** Infrastructure that supports collaboration between research institutions.

    CASE STUDY: Federated Identity at CERN

    CASE STUDY: Nordic e-Infrastructure Collaboration (NeIC)

    CASE STUDY: EGI Federated Cloud

Openness and Community Collaboration

The meaning of openness for OpenStack is prescribed by the Four Opens (*https://governance.openstack.org/tc/reference/opens.html*): open source, open design, open development, and open community. Community development and collaboration is a defining OpenStack value for global science organisations. Scientists, faculty, HPC engineers, operators and developers from academia and leading-edge businesses are fully engaged in the vibrant Scientific OpenStack community.

HPC engineers enjoy contributing to OpenStack's ever-evolving code base to meet stringent workload requirements for all science workloads.

Members of scientific organisations collaborate in various ways:

- OpenStack Operators mailing list (*http://lists.openstack.org/cgi-bin/mailman/listinfo/openstack-operators*), using [scientific-wg] in the subject.
- Internet Relay Chat (IRC) (*https://wiki.openstack.org/wiki/IRC*).
- The OpenStack Scientific Working Group (*https://wiki.openstack.org/wiki/Scientific_working_group*).

The Working Group represents and advances the use cases and needs of research and high-performance computing atop OpenStack. Hundreds of members enjoy this great forum for cross-institutional collaboration, at online and in-person events.

Cambridge University's Journey to the Scientific Working Group

At Cambridge University, world-class research is supported by world-class computing resources. A dedicated and experienced team of architects and administrators work to keep Cambridge's HPC infrastructure at the leading edge.

Research computing itself is broadening and diversifying. Non-traditional HPC requirements such as big data analytics are strengthening. Researchers with different software skill sets and needs are demanding support for new software environments. An accelerating rate of change demands flexibility and agility—uncommon traits in the HPC domain.

OpenStack provides solutions for many of the challenges of delivering flexible infrastructure for research computing. By managing compute resources as an OpenStack private cloud, researchers are able to work in environments tailored to their requirements. The dynamic, automated nature of software-defined infrastructure cuts away time wasted on the distractions of setup, and enables researchers to maximise the time they spend on research itself.

If only it was as simple as that. The journey to using OpenStack in production is not straightforward. While the team at Cambridge University have expertise in designing and managing HPC infrastructure, the skill set for OpenStack is very different. OpenStack is a highly complex system with a steep learning curve. Furthermore, default OpenStack configurations are unlikely to yield optimal performance for research computing environments.

Cambridge's solution strategy was twofold: to build strong relationships with vendors in the OpenStack ecosystem, and to participate actively in the OpenStack community.

One of OpenStack's greatest assets is the welcoming, friendly and helpful community built around it. However, while there were many active users within the OpenStack community with research computing use cases, there was no organised group with the specific interest of supporting this use case.

A critical mass was soon gathered. With assistance from the OpenStack Foundation, the Scientific Working Group was formed, with Cambridge University as one of its co-founders (the other being Monash University in Melbourne, Australia).

The Scientific Working Group exists to support and promote the research computing use case. We have many active members, drawn from a global catchment. Our membership is informal and everybody is welcome. We support one another through sharing information, planning events, and advocacy for research computing in an OpenStack environment. Within a strong and open community, we present the case for a Scientific OpenStack.

OpenStack and Virtualised HPC

Likely doubts over the adoption of OpenStack centre around the impact of infrastructure virtualisation. From the skeptical perspective of an HPC architect, why OpenStack?

- *I have heard the hype*
- *I am skeptical to some degree*
- *I need evidence of benefit*

In this section, we will describe the different forms of overhead that can be introduced by virtualisation, and provide technical details of solutions that mitigate, eliminate or bypass the overheads of software-defined infrastructure.

The Overhead of Virtualisation

Analysis typically shows that the overhead of virtualisation for applications that are CPU-intensive is minimal. The overhead of memory-intensive applications is minimal, provided NUMA configuration is passed through from hypervisor to guest. Similarly, applications that depend on high-bandwidth I/O or network communication for bulk data transfers can achieve levels of performance that are close to equivalent bare metal configurations. Where a significant performance impact is observed, it can often be ascribed to overcommitment of hardware resources or "noisy neighbours"—issues that could equally apply in non-virtualised configurations.

However, there remains a substantial class of applications whose performance is significantly impacted by virtualisation. Some of the causes of that performance overhead are described here.

INCREASED SOFTWARE OVERHEAD ON I/O OPERATIONS

Factors such as storage IOPs and network message latency are often critical for HPC application performance. HPC applications that are sensitive to these factors are poor performers in a conventional virtualised environment. Fully virtualised environments incur additional overhead per I/O operation that can impact performance for applications that depend on such patterns of I/O.

The additional overhead is mitigated through paravirtualisation, in which the guest OS includes support for running within a virtualised environment. The guest OS cooperates with the host OS to improve the overhead of hardware device management. Direct hardware device manipulation is performed in the host OS, keeping the micro-management of hardware closer to the physical device. The hypervisor then presents a more efficient software interface to a simpler driver in the guest OS. Performance improves through streamlining interactions between guest OS and the virtual hardware devices presented to it.

HARDWARE OFFLOAD IN A VIRTUALISED NETWORK

All modern Ethernet NICs provide hardware offload of IP, TCP and other protocols. To varying degrees, these free up CPU cycles from the transformations necessary between data in user buffers and packets on the wire (and vice versa).

In a virtualised environment, the network traffic of a guest VM passes from a virtualised network device into the software-defined network infrastructure running in the hypervisor. Packet processing is usually considerably more complex than a typical HPC configuration. Hardware offload capabilities are often unable to operate or are ineffective in this mode. As a result, networking performance in a virtualised environment can be less performant and more CPU-intensive than an equivalent bare metal environment.

INCREASED JITTER IN VIRTUALISED NETWORK LATENCY

To varying degrees, virtualised environments generate increased system noise effects. These effects result in a longer tail on latency distribution for interrupts and I/O operations.

A bulk synchronous parallel workload, iterating in lock-step, moves at the speed of the slowest worker. If the slowest worker is determined by jitter effects in I/O latency, overall application progress becomes affected by the increased system noise of a virtualised environment.

Using OpenStack to Deliver Virtualised HPC

There is considerable development activity in the area of virtualisation. New levels of performance and capability are continually being introduced at all levels: processor architecture, hypervisor, operating system and cloud orchestration.

BEST PRACTICE FOR VIRTUALISED SYSTEM PERFORMANCE

The twice-yearly cadence of OpenStack software releases leads to rapid development of new capabilities, which improve its performance and flexibility.

Across the OpenStack operators community, there is a continual collaborative process of testing and improvement of hypervisor efficiency. Empirical studies of different configurations of tuning parameters are frequently published and reviewed. Clear improvements are collected into a curated guide on hypervisor performance tuning best practice.

OpenStack Nova compute service supports exposing many hypervisor features for raising virtualised performance. For example:

- Enabling processor architecture extensions for virtualisation.
- Controlling hypervisor techniques for efficiently managing many guests, such as Kernel Same-page Merging (KSM). This can add CPU overhead in return for varying degrees of improvement in memory usage by de-duplicating identical pages. For supporting memory-intensive workloads, KSM can be configured to prevent merging between NUMA nodes. For performance-critical HPC, it can be disabled altogether.
- Pinning virtual cores to physical cores.
- Passing through the NUMA topology of the physical host to the guest enables the guest to perform NUMA-aware memory allocation and task scheduling optimisations.
- Passing through the specific processor model of the physical CPUs can enable use of model-specific architectural extensions and runtime microarchitectural optimisations in high-performance scientific libraries.
- Backing guest memory with huge pages reduces the impact of host Translation Lookaside Buffer (TLB) misses.

By using optimisation techniques such as these, the overhead of virtualisation for CPU-bound and memory-bound workloads is reduced to typically one-two percent of bare metal performance. More information can be found in *Further Reading* for this section, below.

Conversely, by constraining the virtual architecture more narrowly, these tuning parameters make VM migration more difficult in a cloud infrastructure consisting of heterogeneous hypervisor hardware, in particular, this may preclude live migration.

HARDWARE SUPPORT FOR I/O VIRTUALISATION

Hardware devices that support Single-Rooted I/O Virtualization (SR-IOV) enable the hardware resources of the physical function of a device to be presented as many virtual functions. Each of these can be individually configured and passed through into a different VM. In this way, the hardware resources of a network card can provide performance with close to no additional overhead, simultaneously serving the diverse needs of many VMs.

Through direct access to physical hardware, SR-IOV networking places some limitations on software-defined infrastructure. It is not typically possible to apply security group policies to a network interface mapped to an SR-IOV virtual function. This may raise security concerns for externally accessible networks, but should not prevent SR-IOV networking being used internally for high-performance communication between the processes of an OpenStack-hosted parallel workload.

Recent empirical studies have found that using SR-IOV for high-performance networking can reduce the overhead of virtualisation typically to 1-9 percent of bare metal performance for network-bound HPC workloads. Links to examples can be found in *Further Reading* at the end of this section.

USING PHYSICAL DEVICES IN A VIRTUALISED ENVIRONMENT

Some classes of HPC applications make intensive use of hardware acceleration in the form of GPUs, Intel Xeon Phi, etc.

Specialised compute hardware in the form of PCI devices can be included in software-defined infrastructure by pass-through. The device is mapped directly into the device tree of a guest VM, providing that VM with exclusive access to the device.

A virtual machine that makes specific requirements for hardware accelerators can be scheduled to a hypervisor with the resources available, and the VM is "composed" by passing through the hardware it needs from the environment of the host.

The resource management model of GPU devices does not currently adapt to SR-IOV. A GPU device is passed-through to a guest VM in its entirety. A host system with multiple GPUs can pass-through different devices to different systems. Similarly, multiple GPU devices can be passed-through into a single instance and GPUDirect peer-to-peer data transfers can be performed between GPU devices and also with RDMA-capable NICs.

Device pass-through, however, can have a performance impact on virtualised memory management. The IOMMU configuration required for pass-through restricts the use of transparent huge pages. Memory must, therefore, be pinned in a guest VM using pass-through devices. This can limit the flexibility of software-defined infrastructure to over-commit virtualised resources (although over-committed resources are generally unlikely to be worthwhile in an HPC use case). Static huge pages can still be used to provide a boost to virtual memory performance.

The performance overhead of virtualised GPU-intensive scientific workloads has been found to be as little as one percent of bare metal performance. More information can be found in *Further Reading* at the end of this section.

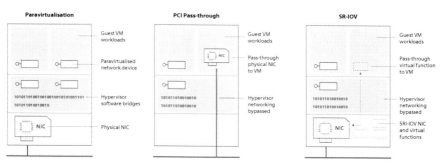

OS-LEVEL VIRTUALISATION: CONTAINERS

Virtualisation overheads are almost eliminated by moving to a different model of compute abstraction. Containers, popularised by Docker, package an application plus its dependencies as a lightweight self-contained execution environment instead of an entire virtual machine. The simplified execution model brings benefits in memory usage and I/O overhead.

Currently, HPC networking using RDMA can be performed within containers, but with limitations. The OFED software stack lacks awareness of network namespaces and cgroups, which prevents per-container control and isolation of RDMA resources. However, containers configured with host networking can use RDMA.

BARE METAL VIRTUALISATION: OPENSTACK IRONIC PROJECT

OpenStack software-defined infrastructure does not need to be virtual.

Ironic is a virtualisation driver. Through some artful abstraction it presents bare metal compute nodes as though they were virtualised compute resources. Ironic's design philosophy results in zero overhead to the performance of the compute node, whilst providing many of the benefits of software-defined infrastructure management.

Through Ironic, a user gains bare metal performance from their compute hardware, but retains the flexibility to run any software image they choose.

The Ironic project is developing rapidly, with new capabilities being introduced with every release. The latest OpenStack release delivers compelling new functionality:

- Serial consoles
- Volume attachment
- Multi-tenant networking

Complex image deployment (over multiple disks, for example) is an evolving capability.

Using Ironic has some limitations:

- Ironic bare metal instances cannot be dynamically intermingled with virtualised instances. However, they can be organised as separate cells or regions within the same OpenStack private cloud.
- Some standard virtualisation features could never be supported, such as overcommitment and migration.

See the section *OpenStack and HPC Infrastructure Management* for further details about Ironic.

Virtualised HPC on OpenStack at Monash University

From its inception in 2012, Australian scientific research has benefited from the NeCTAR Research Cloud federation. Now comprising eight institutions from across the country, NeCTAR was an early adopter of OpenStack, and has been at the forefront of development of the project from that moment.

NeCTAR's federated cloud compute infrastructure supports a wide range of scientific research with diverse requirements. Monash Advanced Research Computing Hybrid (MonARCH) was commissioned in 2015/2016 to provide a flexible and dynamic HPC resource.

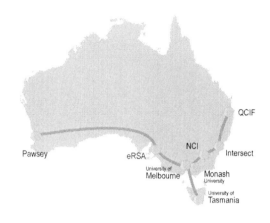

MonARCH has 35 dual-socket Haswell-based compute nodes and 820 CPU cores. MonARCH exploits cloud-bursting techniques to grow elastically by using resources from across the NeCTAR federation. The infrastructure uses a fabric of 56G Mellanox Ethernet for a converged, high-speed network. The cloud control plane is running Ubuntu Trusty and the KVM hypervisor. OpenStack Liberty (as of Q3'2016) was deployed using Ubuntu distribution packages (including selected patches as maintained by NeCTAR Core Services), orchestrated and configured using Puppet.

MonARCH makes extensive use of SR-IOV for accessing its HPC network fabric. The high-speed network is configured to use VLANs for virtual tenant networking, enabling layer-2 RoCEv1 (RDMA over Converged Ethernet). RDMA is used in guest instances in support of tightly coupled parallel MPI workloads, and for high-speed access to 300TB of Lustre storage.

Following MonARCH, Monash University recently built a mixed CPU & GPU cluster called M3, the latest system for the MASSIVE (Multi-modal Australian ScienceS Imaging and Visualisation Environment) project. Within M3, there are 1700 Haswell CPU cores along with 16 quad-GPU compute nodes and an octo-GPU compute node, based upon the NVIDIA K80 dual-GPU. Staff at Monash University's Research @ Cloud Monash (R@CMon) cloud research group have integrated SR-IOV networking and GPU pass-through into their compute instances.

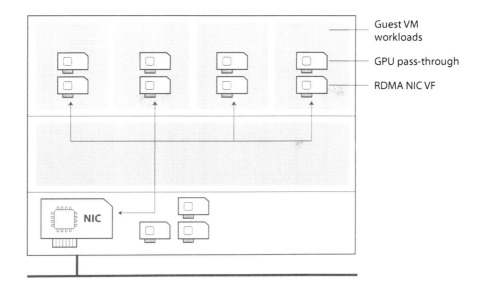

Specific high-performance OpenStack flavors are defined to require pass-through of one or more dedicated GPUs. This enables one to four GPU instances to run concurrently on a dual-K80 compute node, e.g., to support CUDA-accelerated HPC workloads and/or multiple interactive visualisation virtual workstations.

Blair Bethwaite, senior HPC consultant, R@CMon group at Monash University, said:

> *"Using OpenStack brings us a high degree of flexibility in the HPC environment. Applying cloud provisioning and management techniques also helps to make the HPC-stack more generic, manageable and quick to deploy. Plus, we benefit from the constant innovation from the OpenStack community, with the ability to pick and choose new services and projects from the ecosystem. OpenStack flexibility in the SDN space also offers compelling new avenues to integrate researchers' personal or lab servers with the HPC service*
>
> *However, before racing out to procure your next HPC platform driven by OpenStack, I'd recommend evaluating your potential workloads and carefully planning and testing the appropriate mix of hardware capabilities, particularly acceleration features. KVM, OpenStack's most popular hypervisor, can certainly perform adequately for HPC—in recent testing we are getting 98 percent on average and up to 99.9 percent of bare metal in Linpack tests—but a modern*

HPC system is likely to require some subset of bare metal infrastructure. If I was planning a new deployment today, I'd seriously consider including Ironic so that a mix of bare metal and virtual cloud nodes can be provisioned and managed consistently. As Ironic is maturing and becoming more feature-complete, I expect to see many more highly integrated deployments and reference architectures emerging in the years to come."

Optimising for "Time to Paper" using HPC on OpenStack

When evaluating OpenStack as a candidate for HPC infrastructure for research computing, the "time to paper" metric of the scientists using the resource should be included in consideration.

Skeptics of using cloud compute for HPC infrastructure inevitably cite the various overheads of virtualisation in the case against OpenStack. With a rapidly developing technology, these arguments can often be outdated. Furthermore, cloud infrastructure presents a diminishing number of trade-offs in return for an increasing number of compelling new capabilities.

Unlike conventional HPC system management, OpenStack provides, for example:

- *Standardisation* as users can interact with the system through a user-friendly web interface, a command line interface or a software API.
- *Flexibility* and *agility* as users allocate compute resources as required and have exclusive use of the virtual resources. There is fine-grained control of the extent to which physical resources are shared.
- Users can *self-serve* and boot a software image of their choosing without requiring operator assistance. It is even possible for users to create their own software images to run—a powerful advantage that eliminates toil for the administrators and delay for the users.
- Additional *security* as users have a higher degree of separation from each other. They cannot observe other users and are isolated from one another on the network.

Through careful consideration, an HPC-aware configuration of OpenStack is capable of realising all the benefits of software-defined infrastructure whilst incurring minimal overhead. In its various forms, virtualisation strikes a balance between new capabilities and consequential overhead.

Further Reading

The OpenStack Hypervisor Tuning Guide is a living document detailing best practice for virtualised performance: *https://wiki.openstack.org/wiki/Documentation/HypervisorTuningGuide*

CERN's OpenStack in Production blog is a good example of the continual community process of hypervisor tuning: *http://openstack-in-production.blogspot.co.uk/*

As an example of the continuous evolution of hypervisor development, the MIKELANGELO project is currently working on optimisations for reducing the latency of virtualised I/O using their sKVM project: *https://www.mikelangelo-project.eu/2015/10/how-skvm-will-beat-the-io-performance-of-kvm/*

For more information on OpenStack and containers, please visit *https://www.openstack.org/containers/*.

An informative paper describing recent developments enabling GPUDirect peer-to-peer transfers between GPUs and RDMA-enabled NICs: *http://grids.ucs.indiana.edu/ptliupages/publications/15-md-gpudirect%20(3).pdf*

Whilst the focus of this paper is on comparing virtualisation strategies on the ARM architecture, the background information is accessible and the comparisons made with the x86 architecture are insightful: *http://www.cs.columbia.edu/~cdall/pubs/isca2016-dall.pdf*

For more information about MonARCH at Monash University, see the R@CMon blog: *https://rcblog.erc.monash.edu.au/*

OpenStack and HPC Network Fabrics

HPC and enterprise cloud infrastructure are not built to the same requirements. As much as anything else, networking exemplifies the divergent criteria between HPC applications and the typical workloads served by cloud infrastructure.

With sweeping generalisations, one typically assumes an HPC parallel workload is tightly coupled, and a cloud-native workload is loosely coupled. A typical HPC parallel workload might be computational fluid dynamics using a partitioned geometric grid. The application code is likely to be structured in a bulk synchronous parallel model, comprising phases of compute and data exchange between neighbouring workers. Progress is made in lockstep, and is blocked until all workers complete each phase.

Compare this with a typical cloud-native application, which might be a microservice architecture consisting of a number of communicating sequential processes. The overall application structure is completely different, and workers do not have the same degree of dependency upon other workers in order to make progress.

The different requirements of HPC and cloud-native applications have led to different architectural choices being made at every level in order to deliver optimal and cost-effective solutions for each target application.

A cloud environment experiences workload diversity to a far greater extent than seen in HPC. This has become the quintessential driving force of software-defined infrastructure. Cloud environments are designed to be flexible and adaptable. As a result, cloud infrastructure has the flexibility to accommodate HPC requirements.

The flexibility of cloud infrastructure is delivered through layers of abstraction. OpenStack's focus is on defining the intent of the multi-tenant cloud infrastructure. Dedicated network management applications decide on the implementation. As an orchestrator, OpenStack delegates knowledge of physical network connectivity to the network management platforms to which it connects.

OpenStack's surging momentum has ensured that support already exists for all but the most exotic of HPC network architectures. This article will describe several solutions for delivering HPC networking in an OpenStack cloud.

Using SR-IOV for Virtualised HPC Networking

SR-IOV is a technology that demonstrates how software-defined infrastructure can introduce new flexibility in the management of HPC resources, whilst retaining the high-performance benefits. With current generation devices, there is a slight increase in I/O latency when using SR-IOV virtual functions. However, this overhead is negligible for all but the most latency-sensitive of applications.

THE PROCESS FLOW FOR USING SR-IOV

The Nova compute hypervisor is configured at boot time with kernel flags to support extensions for SR-IOV hardware management.

The network kernel device driver is configured to create virtual functions. These are present alongside the physical function. When they are not assigned to a guest workload instance, the virtual functions are visible in the device tree of the hypervisor.

The OpenStack services are configured with identifiers or addresses of devices configured to support SR-IOV. This is most easily done by identifying the physical function (for example using its network device name, PCI bus address, or PCI vendor/device IDs). All virtual functions associated with this device will be made available for virtualised compute instances. The configuration that identifies SR-IOV devices is known as the whitelist.

To use an SR-IOV virtual function for networking in an instance, a special direct-bound network port is created and connected with the VM. This causes one of the virtual functions to be configured and passed-through from the hypervisor into the VM.

Support for launching an instance using SR-IOV network interfaces from OpenStack Horizon web interface was introduced in the OpenStack Mitaka release (April 2016). Prior to this, it was only possible to launch instances using SR-IOV ports through a sequence of command-line invocations (or through direct interaction with the OpenStack APIs).

THE LIMITATIONS OF USING SR-IOV IN CLOUD INFRASTRUCTURE

SR-IOV places some limitations on the cloud computing model that can be detrimental to the overall flexibility of the infrastructure:

- Current SR-IOV hardware implementations support flat (unsegregated) and VLAN network separation but not VXLAN for tenant networks. This limitation can constrain the configuration options for the network fabric. Layer-3 IP-based fabrics using technologies such as ECMP are unlikely to interoperate with VLAN-based network separation.

- Live migration of VMs connected using SR-IOV is not possible with current hardware and software. The capability is being actively developed for Mellanox SR-IOV NICs. It is not confirmed whether live migration of RDMA applications will be possible.
- SR-IOV devices bypass OpenStack's security groups, and consequently should only be used for networks that are not externally connected.

Virtualisation-aware MPI for Tightly Coupled Cloud Workloads

The MVAPICH2 library implements MPI-3 (based on MPI 3.1 standard) using the InfiniBand (IB) verbs low-level message passing primitives. MVAPICH2 was created and developed by the Network-Based Computing Laboratory (*http://nowlab.cse.ohio-state.edu/*) (NOWLAB) at The Ohio State University, and has been freely available for download for 15 years. Over that time MVAPICH2 has been continuously developed and now runs on systems as big as 10,649,600 cores.

An InfiniBand NIC with SR-IOV capability was first developed by Mellanox in the ConnectX-3 generation of its product, unlocking the possibility of achieving near-native InfiniBand performance in a virtualised environment. MVAPICH2-Virt was introduced in 2015 to develop HPC levels of performance for cloud infrastructure. The techniques adopted by MVAPICH2-Virt currently support KVM- and Docker-based cloud environments. MVAPICH2-Virt introduces Inter-VM Shared Memory (IVSHMEM) support to KVM hypervisors, increasing performance between co-resident VMs. In order to run MVAPICH2-Virt-based applications on top of OpenStack-based cloud environments easily, several extensions to set up SR-IOV and IVSHMEM devices in VMs have been developed for OpenStack Nova compute manager.

MVAPICH2-Virt has two principal optimisation strategies for KVM-based cloud environments:

- Dynamic locality awareness for MPI communication among co-resident VMs. A new communication channel, IVSHMEM, introduces a memory-space communication mechanism between different VMs co-resident on the same hypervisor. Inter-node communication continues to use the SR-IOV virtual function.
- Tuning of MPI performance for both SR-IOV and IVSHMEM channels.

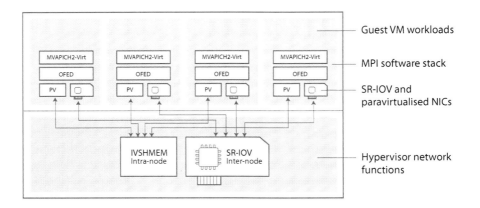

Similarly, MVAPICH2-Virt has two principal optimisation strategies for Docker-based cloud environments:

- Dynamic locality awareness for MPI communication among co-resident containers. All Intra-Node MPI communication can go through either an IPC-SHM-enabled channel or CMA channel, no matter if they are in the same container or different ones. Inter-Node-Inter-Container MPI communication will leverage the InfiniBand HCA channel.
- Tuning of MPI performance for all different channels, including IPC-SHM, CMA, and InfiniBand HCA.

With these strategies in effect, the performance overhead of KVM- and Docker-based virtualisation on standard MPI benchmarks and applications are less than ten percent.

> "The novel designs introduced in MVAPICH2-Virt take advantage of the latest advances in virtualisation technologies and promise to design next-generation HPC cloud environments with good performance." —Prof. DK Panda and Dr. Xiaoyi Lu of NOWLAB

OpenStack as an environment for supporting MPI-based HPC workloads has many benefits, such as fast VM or container deployment for setting up MPI job execution environments, security, enabling resource sharing, providing privileged access in virtualised environments, supporting high-performance networking technologies (e.g., SR-IOV), etc.

Currently, OpenStack still could not fully support or work seamlessly with technologies in HPC environments, such as IVSHMEM, Slurm, PBS, etc. But with several extensions proposed by NOWLAB researchers, running MPI-based HPC workloads on top of OpenStack-managed environments seems a promising approach for building efficient clouds.

The future direction of MVAPICH2-Virt includes:

- Further support different kinds of virtualised environments.
- Further improve MPI application performance on cloud environments through novel designs.
- Support live migration of MPI applications in SR-IOV- and IVSHMEM-enabled VMs.

InfiniBand and other Non-Ethernet Fabrics

InfiniBand is the dominant fabric interconnect for HPC clusters. Of the TOP500 list published in June 2017 (*https://www.top500.org/lists/2017/06/*), 36 percent of entries use InfiniBand.

In part, OpenStack's flexibility comes from avoiding many rigid assumptions in infrastructure management. However, OpenStack networking does have some expectations of an Ethernet and IP-centric network architecture, which can present challenges for the network architectures often used in HPC. The Neutron driver for InfiniBand circumvents this assumption by applying Neutron's layer-3 network configuration to an IP-over-IB interface, and mapping Neutron's layer-2 network segmentation ID to InfiniBand pkeys.

However, Neutron is limited to an allocation of 126 pkeys, which imposes a restrictive upper limit on the number of distinct tenant networks an OpenStack InfiniBand cloud can support.

A technical lead with experience of using OpenStack on InfiniBand reports mixed experiences from an evaluation performed in 2015. The overall result led him to conclude that HPC fabrics such as InfiniBand are only worthwhile in an OpenStack environment if one is also using RDMA communication protocols in the client workload:

> *"MPI jobs were never a targeted application for our system. Rather, the goal for our OpenStack was to accommodate all the scientific audiences for whom big HPC clusters, and batch job schedulers, weren't a best fit. So, no hard, fast requirement for a low-latency medium. What we realised was that it's complex. It may be hard to keep it running in production. IPoIB on FDR, in unconnected mode, is slower than 10Gbps Ethernet, and if you're not making use of RDMA, then you're really just kind of hurting yourself. Getting data in and out is tricky. All the big data we have is on a physically separate IB fabric, and no one wanted to span those fabrics, and doing something involving IP routing would break down the usefulness of RDMA."*

IP-over-IB performance and scalability has improved substantially in subsequent hardware and software releases. A modern InfiniBand host channel adaptor with a current driver stack operating in connected mode can sustain 35-40 Gbits/sec in a single TCP stream on FDR InfiniBand.

The Canadian HPC4Health consortium has deployed a federation of OpenStack private clouds using a Mellanox FDR InfiniBand network fabric.

Intel Omnipath network architecture is starting to emerge in the Scientific OpenStack community. At Pittsburgh Supercomputer Center, the Bridges system entered production in early 2016 for HPC and data analytics workloads. It comprises over 800 compute nodes with an Omnipath fabric interconnect. In its current product generation, Omnipath does not support SR-IOV. Bridges is a bare metal system, managed using OpenStack Ironic. The Omnipath network management is managed independently of OpenStack.

Bridges is described in greater detail in the *OpenStack and HPC Infrastructure Management* section.

An RDMA-Centric Bioinformatics Cloud at Cambridge University

Cambridge University's Research Computing Services group has a long track record as a user of RDMA technologies such as InfiniBand and Lustre across all its HPC infrastructure platforms. When scoping a new bioinformatics compute resource in 2015, the desire to combine this proven HPC technology with a flexible self-service platform led to a requirements specification for an RDMA-centric OpenStack cloud.

Bioinformatics workloads can be I/O-intensive in nature, and can also feature I/O access patterns that are highly sensitive to I/O latency. Whilst this class of workload is typically a weakness of virtualised infrastructure, the effects are mitigated through use of HPC technologies such as RDMA and virtualisation technologies such as SR-IOV to maximise efficiency and minimise overhead.

The added complexity of introducing HPC networking technologies is considerable, but remains hidden from the bioinformatics users of the system. Block-based I/O via RDMA is delivered to the kernel of the KVM hypervisor. The compute instances simply see a paravirtualised block device. File-based I/O via RDMA is delivered using the Lustre filesystem client drivers running in the VM instances. Through use of SR-IOV virtual functions, this is identical to a bare metal compute node in a conventional HPC configuration. Similarly, MPI communication is performed on the virtualised network interfaces with no discernable difference for the user of the compute instance.

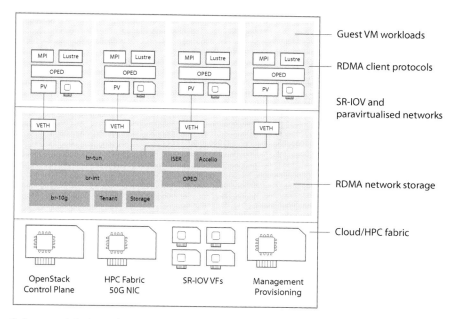

Software architecture of the compute node of an RDMA-centric OpenStack cloud

The cloud contains 80 compute nodes, three management nodes and a number of storage nodes of various kinds. The system runs Red Hat OpenStack Platform (OSP) and is deployed using Red Hat's TripleO-based process. All the HPC-centric features of the system have been implemented using custom configuration and extensions to TripleO. Post-deployment configuration management is performed using Ansible-OpenStack playbooks, resulting in a devops approach for managing an HPC system.

To deploy a system with RDMA networking enabled in the compute node hypervisor, overcloud management QCOW2 images are created with OpenFabrics installed. Cinder is configured to use iSER (iSCSI Extensions for RDMA) as a transport protocol.

The cloud uses a combination of Mellanox 50G Ethernet NICs and 100G Ethernet switches for its HPC network fabric. RDMA support using RoCEv1 requires layer-2 network connectivity. Consequently, OpenStack networking is configured to use VLANs for control plane traffic and HPC tenant networking. VXLAN is used for other classes of tenant networking.

A multi-path layer-2 network fabric is created using multi-chassis LAGs. Traffic is distributed across multiple physical links whilst presenting a single logical link for the Ethernet network topology. Port memberships of the tenant network VLANs are managed dynamically using the NEO network management platform from Mellanox, which integrates with OpenStack Neutron.

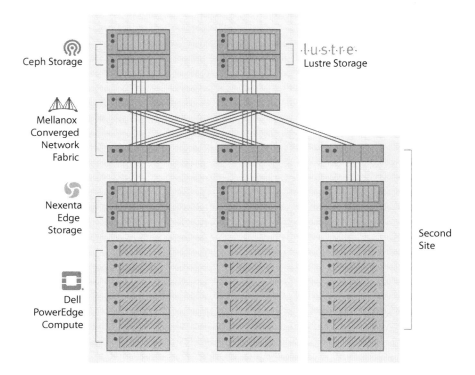

The Forces Driving HPC and Cloud Diverge in Network Management

At the pinnacle of HPC, ultimate performance is achieved through exploiting full knowledge of all hardware details: the microarchitecture of a processor, the I/O subsystem of a server—or the physical location within a network. HPC network management delivers performance by enabling workload placement with awareness of the network topology.

The cloud model succeeds because of its abstraction. Cloud infrastructure commits to delivering a virtualised flat network to its instances. All details of the underlying physical topology are obscured. Where an HPC network management system can struggle to handle changes in physical network topology, cloud infrastructure adapts.

OpenStack provides a limited solution to locality-aware placement, through use of Availability Zones (AZ). By defining an AZ per top-of-rack switch, a user can request that instances be scheduled to be co-resident on the same edge switch. However, this can be a clumsy interface for launching instances on a large private cloud, and AZs cannot be nested to provide multiple levels of locality for co-locating larger workloads.

OpenStack depends on other network management platforms for physical network knowledge, and delegates to them all aspects of physical network management. Network management and monitoring packages such as LibreNMS, Observium and Mellanox NEO are complementary to the functionality provided by OpenStack.

Another key theme in HPC network management is in gathering network-centric performance telemetry.

While HPC does not deliver on all of its promises in this area, there is greater focus within HPC network management on the ability to collect telemetry data on the performance of a network for optimising the workload.

Business-oriented clouds and HPC take very different approaches in this sector.

In general, HPC performance monitoring is done at the application level. HPC application performance analysis typically follows a model in which runtime trace data is gathered during execution for later aggregation and visualisation. This approach avoids overhead when monitoring is not required and minimises the overhead when monitoring is active. When application monitoring is active, leading packages such as OVIS minimise overhead by using RDMA for aggregation of runtime telemetry data. Application performance visualisation is performed using tools such as VAMPIR. All these HPC-derived application performance monitoring tools will also work for applications running within an OpenStack/HPC environment.

At a system level, HPC network performance analysis is more limited in scope, but developments such as PAVE at Lawrence Livermore and more recently INAM[2] from Ohio State University are able to demonstrate a more holistic capability to identify adverse interactions between applications sharing a network, in addition to performance bottlenecks within an application itself.

The pace of development of cloud infrastructure monitoring is faster, and in many cases is derived from open-source equivalents of hyperscaler-developed capabilities. Twitter's Zipkin is a distributed application performance monitoring framework derived from conceptual details from Google's Dapper. LinkedIn developed and published Kafka, a distributed near-real-time message log. However, the layers of abstraction that give cloud its flexibility can prevent cloud monitoring from providing performance insights from the physical domain that inform performance in the virtual domain.

At the OpenStack Summit in Paris in November 2014, Intel demonstrated Apex Lake, a project which aims to provide performance telemetry across these abstraction boundaries—including across virtual/physical network abstractions. Some of these features may have been incorporated into the Intel's open source Snap telemetry/monitoring framework.

In its present situation, through use of SR-IOV network devices, cloud network infrastructure has demonstrated that it is capable at achieving performance levels that are typically within one to nine percent of bare metal. OpenStack can be viewed as the integration and orchestration of existing technology platforms. The physical network performance telemetry of cloud network infrastructure is delegated to the technology platforms upon which it is built. In future, projects such as INAM[2] on the HPC side and Apex Lake on the cloud side may lead to a telemetry monitoring framework capable of presenting performance data from virtual and physical domains in the context of one another.

Further Reading

This Intel white paper provides a useful introduction to SR-IOV: *https://www.intel.com/content/dam/doc/white-paper/pci-sig-single-root-io-virtualization-support-in-virtualization-technology-for-connectivity-paper.pdf*

A step-by-step guide to setting up Mellanox InfiniBand with a Red Hat variant of Linux[3] and OpenStack Mitaka: *https://wiki.openstack.org/wiki/Mellanox-Neutron-Mitaka-Redhat-InfiniBand*

A presentation by Professor DK Panda and Dr. Xiaoyi Lu from NOWLAB at The Ohio State University on MVAPICH2-Virt: *https://www.openstack.org/videos/austin-2016/building-efficient-hpc-clouds-with-mvapich2-and-openstack-over-sr-iov-enabled-InfiniBand-clusters*

Further information on MVAPICH2-Virt can be found here: *http://mvapich.cse.ohio-state.edu*

Papers from the team at NOWLAB describing MVAPICH2-Virt in greater depth:
- [HiPC'14] High Performance MPI Library over SR-IOV Enabled InfiniBand Clusters
  Jie Zhang, Xiaoyi Lu, Jithin Jose, Rong Shi, Mingzhe Li, and Dhabaleswar K. (DK) Panda.
  Proceedings of the 21st annual IEEE International Conference on High Performance Computing (HiPC), 2014.
- [CCGrid'15] MVAPICH2 over OpenStack with SR-IOV: An Efficient Approach to Build HPC Clouds
  Jie Zhang, Xiaoyi Lu, Mark Arnold, and Dhabaleswar K. (DK) Panda.
  Proceedings of the 15th IEEE/ACM International Symposium on Cluster, Cloud and Grid Computing (CCGrid), 2015.

The OVIS HPC application performance monitoring framework: *https://ovis.ca.sandia.gov/mediawiki/index.php/Main_Page*

PAVE – Performance Analysis and Visualisation at Exascale at Lawrence Livermore: *http://computation.llnl.gov/projects/pave-performance-analysis-visualization-exascale*

The introduction of INAM[2] for real-time InfiniBand network performance monitoring: *http://mvapich.cse.ohio-state.edu/static/media/publications/abstract/subramoni-isc16-inam.pdf*

Further information on INAM[2] can be found here: *http://mvapich.cse.ohio-state.edu/tools/osu-inam/*

Open Zipkin is a distributed application performance monitoring framework developed at Twitter. Zipkin is based on the Google Dapper monitoring framework paper: *http://zipkin.io*

Intel Snap is a new monitoring framework for virtualised infrastructure: *http://snap-telemetry.io*

LibreNMS is a network mapping and monitoring platform built upon SNMP: *https://www.librenms.org/*

A useful discussion on the value of high-resolution network telemetry for researching issues with maximum latency in a cloud environment: *https://engineering.linkedin.com/performance/who-moved-my-99th-percentile-latency*

OpenStack and High-Performance Data

What can data requirements mean in an HPC context? The range of use cases is almost boundless. With considerable generalisation we can consider some broad criteria for requirements, which expose the inherent tensions between HPC-centric and cloud-centric storage offerings:

- The *data access* model: data objects could be stored and retrieved using file-based, block-based, object-based or stream-based access. HPC storage tends to focus on a model of file-based shared data storage. Object-based storage is more strongly favoured in cloud infrastructure, although there is an emerging trend for promoting object storage for exascale HPC (DAOS, for example). Conversely, cloud infrastructure favours block-based storage models, often backed with and extended by object-based storage. Support for data storage through shared filesystems is still maturing in OpenStack.

- The *data sharing* model: applications may request the same data from many clients, or the clients may make data accesses that are segregated from one another. This distinction can have significant consequences for storage architecture. Cloud storage and HPC storage are both highly distributed, but often differ in the way in which data access is parallelised. Providing high-performance access for many clients to a shared dataset can be a niche requirement specific to HPC. Cloud-centric storage architectures typically focus on delivering high aggregate throughput on many discrete data accesses.

- The level of *data persistence*. An HPC-style tiered data storage architecture does not need to incorporate data redundancy at every level of the hierarchy. This can improve performance for tiers caching data closer to the processor.

The cloud model offers capabilities that enable new possibilities for HPC:

- *Automated provisioning.* Software-defined infrastructure automates the provisioning and configuration of compute resources, including storage. Users and group administrators are able to create and configure storage resources to their specific requirements at the exact time they are needed.

- *Multi-tenancy.* HPC storage does not offer multi-tenancy with the level of segregation that cloud can provide. A virtualised storage resource can be reserved for the private use of a single user, or could be shared between a controlled group of collaborating users, or could even be accessible by all users.

- *Data isolation.* Sensitive data requires careful data management. Medical informatics workloads may contain patient genomes. Engineering simulations may contain data that is a trade secret. OpenStack's segregation model is stronger than ownership and permissions on a POSIX-compliant shared file system, and also provides finer-grained access control.

There is clear value in increased flexibility—but at what cost in performance? In more demanding environments, HPC storage tends to focus on and be tuned for delivering the requirements of a confined subset of workloads. This is the opposite approach to the enterprise cloud model, in which assumptions may not be possible about the storage access patterns of the supported workloads.

This study will describe some of these divergences in greater detail, and demonstrate how OpenStack can integrate with HPC storage infrastructure. Finally, some methods of achieving high-performance data management on cloud-native storage infrastructure will be discussed.

File-based Data: HPC Parallel Filesystems in OpenStack

Conventionally in HPC, file-based data services are delivered by parallel filesystems such as Lustre and IBM Spectrum Scale–General Parallel File System (GPFS). A parallel file system is a shared resource. Typically it is mounted on all compute nodes in a system and available to all users of a system. Parallel filesystems excel at providing low-latency, high-bandwidth access to data.

Parallel filesystems can be integrated into an OpenStack environment in a variety of configuration models.

PROVISIONED CLIENT MODEL

Access to an external parallel file system is provided through an OpenStack provider network. OpenStack compute instances—virtualised or bare metal—mount the site filesystem as clients.

This use case is fairly well established. In the virtualised use case, performance is achieved through use of SR-IOV (with only a moderate level of overhead). In the case of Lustre, with a layer-2 VLAN provider network the o2ib client drivers can use RoCE to perform Lustre data transport using RDMA.

Cloud-hosted clients on a parallel filesystem raise issues with pervasive use of root privileges in a cloud compute context. On cloud infrastructure, privileged accesses from a client do not have the same degree of trust as on conventional HPC infrastructure. Lustre approaches this issue by introducing Kerberos authentication for filesystem mounts and subsequent file accesses. Kerberos credentials for Lustre filesystems can be supplied to OpenStack instances upon creation as instance metadata.

PROVISIONED FILE SYSTEM MODEL

There are use cases where the dynamic provisioning of software-defined parallel filesystems has considerable appeal. There have been proof-of-concept demonstrations of provisioning Lustre filesystems from scratch using OpenStack compute, storage and network resources.

The OpenStack Manila project aims to provision and manage shared filesystems as an OpenStack service. Spectrum Scale integrates with Manila to re-export GPFS parallel filesystems using the user-space Ganesha NFS server.

Currently these projects demonstrate functionality over performance. In future evolutions, the overhead of dynamically provisioned parallel filesystems on OpenStack infrastructure may be reduced.

A PARALLEL DATA SUBSTRATE FOR OPENSTACK SERVICES

IBM positions Spectrum Scale as a distributed data service for underpinning OpenStack services such as Cinder, Glance, Swift and Manila. More information about using Spectrum Scale in this manner can be found in the IBM Research red paper on the subject, listed in *Further Reading* for this section.

Applying HPC Technologies to Enhance Data I/O

A recurring theme throughout this study has been the use of remote DMA for efficient data transfer in HPC environments. The advantages of this technology are especially pertinent in data-intensive environments. OpenStack's flexibility enables the introduction of RDMA protocols for many cloud infrastructure operations to reduce latency, increase bandwidth and enhance processor efficiency:

- Cinder block data I/O can be performed using iSER. iSER is a drop-in replacement for iSCSI that is easy to configure and set up. Through providing tightly coupled I/O resources using RDMA technologies, the functional equivalent of HPC-style burst buffers can be added to the storage tiers of cloud infrastructure.

- Ceph data transfers can be performed using the Accelio RDMA transport. This technology was demonstrated some years ago but does not appear to have achieved production levels of stability or gained significant mainstream adoption.

- The NOWLAB group at The Ohio State University has developed extensions to data analytics platforms such as HBase, Hadoop, Spark and Memcached to optimise data movements using RDMA.

Optimising Ceph Storage for Data-Intensive Workloads

The versatility of Ceph embodies the cloud-native approach to storage, and consequently Ceph has become a popular choice of storage technology for OpenStack infrastructure. A single Ceph deployment can support various protocols and data access models.

Ceph is capable of delivering strong read bandwidth. For large reads from OpenStack block devices, Ceph is able to parallelise the delivery of the read data across multiple OSDs.

Ceph's data consistency model commits writes to multiple OSDs before a write transaction is completed. By default, a write is replicated three times. This can result in higher latency and lower performance on write bandwidth. Through configuring Ceph pools with erasure coding instead of replication, greater efficiency can be achieved for usable storage capacity.

Ceph can run on clusters of commodity hardware. However, in order to maximise the performance (or price performance) of a Ceph cluster, some design rules of thumb can be applied:

- Use separate physical network interfaces for external storage network and internal storage management. On the NICs and switches, enable Ethernet flow control and raise the MTU to support jumbo frames.
- Each drive used for Ceph storage is managed by an OSD process. A Ceph storage node usually contains multiple drives (and multiple OSD processes).
- The best price/performance and highest density is achieved using fat storage nodes, typically containing 72 HDDs. These work well for large-scale deployments, but can lead to very costly units of failure in smaller deployments. Node configurations of 12-32 HDDs are usually found in deployments of intermediate scale.
- Ceph storage nodes usually contain a higher-speed write journal, which is dedicated to service of a number of HDDs. An SSD journal can typically feed six HDDs while an NVMe flash device can typically feed up to 20 HDDs.
- About 10G of external storage network bandwidth balances the read bandwidth of up to 15 HDDs. The internal storage management network should be similarly scaled.
- A rule of thumb for RAM is to provide 0.5GB–1GB of RAM per TB per OSD daemon.
- On multi-socket storage nodes, close attention should be paid to NUMA considerations. The PCI storage devices attached to each socket should be working together. Journal devices should be connected with HDDs attached

- to HBAs on the same socket. IRQ affinity should be confined to cores on the same socket. Associated OSD processes should be pinned to the same cores.
- For tiered storage applications in which data can be regenerated from other storage, the replication count can safely be reduced from three to two copies.

The Cancer Genome Collaboratory: Large-scale Genomics on OpenStack

Genome datasets can be hundreds of terabytes in size, sometimes requiring weeks or months to download and significant resources to store and process.

The Ontario Institute for Cancer Research (OICR) built Cancer Genome Collaboratory (or simply Collaboratory) as a biomedical research resource built upon OpenStack infrastructure. Collaboratory aims to facilitate research on the world's largest and most comprehensive cancer genome dataset, currently produced by the International Cancer Genome Consortium (ICGC) (*http://icgc.org/*).

By making the ICGC data available in cloud compute form in the Collaboratory, researchers can bring their analysis methods to the cloud, yielding benefits from the high availability, scalability and economy offered by OpenStack, and avoiding the large investment in compute resources and the time needed to download the data.

AN OPENSTACK ARCHITECTURE FOR GENOMICS

Collaboratory's requirements for the project were to build a cloud-computing environment providing 3000 compute cores and 10-15 PB of raw data stored in a scalable and highly available manner. The project has also met constraints of budget, sustainability, data security, confined data centre space, power and connectivity. In selecting the storage architecture, capacity was considered to be more important than latency and performance.

Each rack hosts 16 compute nodes using 2U high-density chassis, and between 6 and 8 Ceph storage nodes. Hosting a mix of compute and storage nodes in each rack keeps some of the Nova-Ceph traffic in the same rack, while also lowering the power requirement for these high-density racks (2 x 60A circuits are provided to each rack).

As of September 2017, Collaboratory has 72 compute nodes (2600 CPU cores, hyper-threaded) with a physical configuration optimised for large data-intensive workflows: 32 or 40 CPU cores and a large amount of RAM (256GB per node). The workloads make extensive use of high-performance local disk, incorporating hardware RAID10 across 6 x 2TB SAS drives.

The networking is provided by Brocade ICX 7750-48C top-of-rack switches that use 6x40GB cables to interconnect the racks in a ring stack topology, providing 240 Gbps non-blocking redundant inter-rack connectivity, at a 2:1 oversubscription ratio.

Collaboratory is deployed using entirely community-supported free software. The OpenStack control plane is Ubuntu 16.04 and deployment configuration is based on custom-developed Ansible playbooks. Collaboratory was initially deployed using OpenStack Juno and has since had upgrades through Kilo, Liberty, Mitaka and Newton.

Collaboratory deploys a standard HA stack based on HAProxy/Keepalived and MariaDB-Galera using three controller nodes. The controller nodes also perform the role of ceph-mon and Neutron L3-agents, using separate partitions for MySQL and ceph-mon on a RAID10 of six SSD drives.

The compute nodes have 10G Ethernet with GRE and SDN capabilities for virtualised networking. The Ceph nodes use 2x10G NICs bonded for client traffic and 2x10G NICs bonded for storage replication traffic. The controller nodes have 4x10G NICs in an active-active bond (802.3ad) using layer3+4 hashing for better link utilisation. The OpenStack tenant routers are highly available with two routers distributed across the three controllers. The configuration does not use Neutron DVR out of a wish to limit the number of servers directly attached to the Internet. The public VLAN is carried only on the trunk ports facing the controllers and the monitoring server.

OPTIMISING CEPH FOR GENOMICS WORKLOADS

Upon workload start, the instances usually download data stored in Ceph object storage. OICR developed a download client (*http://docs.icgc.org/cloud/guide/*) that controls access to sensitive ICGC protected data through managed tokens. Downloading a 100GB file stored in Ceph takes around 18 minutes, with another 10-12 minutes used to automatically check its integrity (md5sum), and is mostly limited by the instance's local disk speed.

The ICGC storage system adds a layer of control on top of Ceph object storage. Currently, this is a two-node cluster behind an HAProxy instance serving the ICGC storage client. The server component uses OICR authorisation and metadata systems to provide secure access to related objects stored in Ceph. By using OAuth-based access tokens, researchers can be given access to the Ceph data without having to configure Ceph permissions. Access to individual project groups can also be implemented in this layer.

Collaboratory currently runs Ceph Jewel. Each Ceph storage node consists of 36 OSD drives (4, 6, 8, or 10TB) in a large Ceph cluster currently providing 7.6PB of raw storage, using three replica pools. The radosgw pool has 90 percent of the Ceph space being reserved for storing protected ICGC datasets, including the very large whole genome aligned reads for almost 2,000 donors. The remaining 10 percent of Ceph space is used as a scalable and highly-available backend for Glance and Cinder. Ceph radosgw was tuned for the specific genomic workloads, mostly by increasing read-ahead on the OSD nodes, 65GB as RADOS (Reliable Autonomic Distributed Object Store) object stripe for radosgw and 8MB for RBD.

A SUSTAINABILITY MODEL FOR RESEARCH CLOUDS

Collaboratory started cost recovery in August 2017, after implementing a billing and reporting application. This in-house developed solution, at the cornerstone of the project's sustainability model, allows the project to provision for hardware renewal by issuing monthly invoices corresponding to actual usage.

Since Collaboratory is a not-for-profit initiative, all of the collected funds are directly re-invested into the project. With more than 30 research projects, from four continents, and a growing user base, the team hopes to sustain, and eventually grow the infrastructure while remaining extremely competitive compared to commercial cloud providers.

The team is building an enrollment application to allow principal investigators to submit new project requests, as well as administrate their projects (in particular, add/remove users request). The new application will streamline and simplify the onboarding process and is expected to be rolled out in 3Q 2017.

All of the components implemented by the team as part of this model are open source and have been containerised for ease of integration by other OpenStack-based clouds.

FURTHER CONSIDERATIONS AND FUTURE DIRECTIONS

In the course of the development of the OpenStack infrastructure at Collaboratory, several issues have been encountered and addressed:

- The instances used in cancer research are usually short lived (hours/days/weeks), but with high resource requirements in terms of CPU cores, memory and disk allocation. As a consequence of this pattern of usage, the Collaboratory OpenStack infrastructure does not support live migration as a standard operating procedure.

- Collaboratory has encountered a few problems caused by radosgw bugs involving overlapping multipart uploads. However, these were detected by the Collaboratory monitoring system, and did not result in data loss. The Collaboratory created a monitoring system that uses automated Rally tests to monitor end-to-end functionality, and also downloads a random large S3 object (around 100GB) to confirm data integrity and monitor object storage performance.

- Because of the mix of very large (BAM), medium (VCF) and very small (XML, JSON) files, the Ceph OSD nodes have imbalanced load; the support team has to regularly monitor and rebalance data.

Currently, Collaboratory is hosting 600TB of data from 2,000 donors. Over the next year, OICR will increase the number of ICGC genomes available in Collaboratory, with the goal of having the entire ICGC data set of 25,000 donors estimated to be 5PB when the project completes in 2018.

Although Collaboratory started accepting new projects only in the spring of 2017, there were more than 1,500 instances started in the last 6 months. One project that used Collaboratory heavily was the PanCancer Analysis of Whole Genomes (PCAWG), which characterized the somatic and germline variants from over 2,800 ICGC cancer whole genomes in 20 primary tumour sites.

In conclusion, the Collaboratory environment has been running well for OICR and its partners. George Mihaiescu, senior cloud architect at OICR, has many future plans for OpenStack and the Collaboratory:

> *"We hope to add new OpenStack projects to the Collaboratory offering of services, with Magnum and Mistral as the first candidates. We would also like to provide new compute node configurations with RAID0 instead of RAID10, or even SSD-based local storage for improved I/O performance. Semi-annual OpenStack upgrades and keeping up with the Ceph LTS cycle are also must-do items on our priority list."*

CLIMB: OpenStack, Parallel Filesystems and Microbial Bioinformatics

The CLoud Infrastructure for Microbial Bioinformatics (CLIMB) is a collaboration between four UK universities (Swansea, Warwick, Cardiff and Birmingham) and funded by the UK's Medical Research Council. CLIMB provides compute and storage as a free service to academic microbiologists in the UK. After an extended period of testing, the CLIMB service was formally launched in July 2016.

CLIMB is a federation of four sites, configured as OpenStack regions. Each site has an approximately equivalent configuration of compute nodes, network and storage.

The compute node hardware configuration is tailored to support the memory-intensive demands of bioinformatics workloads. The system as a whole comprises 7680 CPU cores, in fat 4-socket compute nodes with 512GB RAM. Each site also has three large memory nodes with 3TB of RAM and 192 hyper-threaded cores.

The infrastructure is managed and deployed using xCAT cluster management software. The system runs the Kilo release of OpenStack, with packages from RDO (Red Hat Distribution of OpenStack; a freely available packaging of OpenStack for Red Hat systems). Configuration management is automated using Salt.

Each site has 500TB of GPFS storage. Every hypervisor is a GPFS client, and uses an InfiniBand fabric to access the GPFS filesystem. GPFS is used for scratch storage space in the hypervisors.

For longer term data storage, to share datasets and VMs, and to provide block storage for running VMs, CLIMB deploys a storage solution based on Ceph. The Ceph storage is replicated between sites. Each site has 27 Dell R730XD nodes for Ceph storage servers. Each storage server contains 16 4TB HDDs for Ceph OSDs, giving a total raw storage capacity of 6912TB. After three-way replication, this yields a usable capacity of 2304TB.

On two sites, Ceph is used as the storage backend for Swift, Cinder and Glance. At Birmingham, GPFS is used for Cinder and Glance, with plans to migrate to Ceph.

In addition to the InfiniBand network, a Brocade 10G Ethernet fabric is used, in conjunction with dual-redundant Brocade Vyatta virtual routers to manage cross-site connectivity.

In the course of deploying and trialling the CLIMB system, a number of issues have been encountered and overcome:

- The Vyatta software routers were initially underperforming with consequential impact on inter-site bandwidth.
- Some performance issues have been encountered due to NUMA topology awareness not being passed through to VMs.
- Stability problems with Broadcom 10GBaseT drivers in the controllers led to reliability issues. (Thankfully, the HA failover mechanisms were found to work as required.)
- Problems with interactions between Ceph and Dell hardware RAID cards.
- Issues with InfiniBand and GPFS configuration.

CLIMB has future plans for developing their OpenStack infrastructure, including:

- Migrating from regions to Nova cells as the federation model between sites.
- Integrating OpenStack Manila for exporting shared filesystems from GPFS to guest VMs.

Secure Lustre on OpenStack at the Sanger Institute

BACKGROUND AND GOALS

Current scientific software and HPC applications rely heavily upon performant, shared POSIX-compliant filesystems. The combination of ever larger datasets, the personal information that they may contain, and evolving privacy laws mean that it is more challenging than ever to meet the legislative requirements and high performance access to data at scale, which makes scientific research possible.

The bioinformatic pipelines at Sanger are no exception. Addressing large-scale scientific computation challenges with appropriate data access and cross-group restrictions requires solutions that can be applied to both current and developing IT and scientific instrument technologies. There is a clear requirement for a performant, multi-tenant, high-performance clustered filesystem with a relatively low barrier to entry and using existing filesystem features, wherever possible.

SYSTEM CONFIGURATION

The Sanger Institute deployed OpenStack using Red Hat OpenStack Platform 8, which is based on the upstream Liberty release. The hardware platform is SuperMicro, with ~5500 CPU cores available to users, and 4PB of usable Ceph storage providing block (Cinder) and S3-compatible (radosgw) access. Networking for OpenStack is provided by Arista switches in a leaf-and-spine arrangement, allowing for more east/west traffic than the layouts used for our traditional high throughput clusters. Bonding/port-channel is used to provide high availability server connections. Compute nodes (hypervisors) are connected at 2x25GbE; Ceph and controller nodes have 2x100GbE.

LUSTRE ADVANTAGES AND DRAWBACKS

Lustre is a high-performance parallel POSIX filesystem which has been in use at the Sanger Institute for many years. The filesystem capacity in use exceeds 13PB and is currently provided by DDN. Clients are all Ethernet-connected, with InfiniBand used for the backend. In the most recent installations, each Lustre component server typically has 40GbE connectivity. Many of the tools and pipelines in use build on the expectation of having a fast shared filesystem. Lustre's main drawback in presenting data to OpenStack clients is that file permissions are enforced on the client side. With instance owners having "root" access, that security is easily bypassed with a simple "su" command.

INTEGRATION SOLUTION

With help from DDN staff, we were able to provide secure, isolated multi-tenant access from user-controlled OpenStack instances to Lustre filesystems by combining features of both systems. The configuration does not require complex configuration of the clients (e.g. Kerberos).

On the networking side, provider networks (VLANs created outside OpenStack) provide the connectivity; while custom policy rules in Neutron enable access control, restricting which users can connect instances to which provider networks.

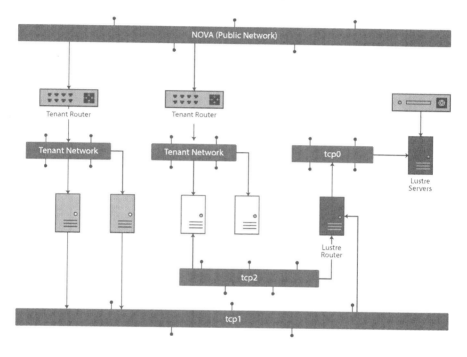

Centrally-controlled Lustre routers were configured to connect provider networks to the Lustre servers, only allowing access to the appropriate provider network IP ranges. Finally, the nodemap feature in Lustre 2.9 onward provides UID/GID mapping, such that whatever accounts are configured within user-controlled instances, the file accesses are "squashed" to an administrator-controlled UID/GID on the filesystem and subtree mounts. This ensures each tenant is constrained to a subdirectory of the filesystem and cannot access data outside that.

FUTURE WORK

Some of the configuration is amenable to automation and we have partially implemented this with Ansible. However, creating new roles currently requires Neutron to be restarted, which can be disruptive. We mitigate this by creating batches of networks/roles in advance.

Where OpenStack instances are under centralised control, the UID/GID mapping and subtree constraints are unnecessary; a provider network can be used to give NAT-free connectivity to a Lustre filesystem. We have prototyped adding virtual compute nodes to an LSF cluster alongside physical compute nodes. Users see identical access to the Lustre filesystems, and the virtual compute nodes are managed and configured by Ansible in exactly the same way as the physical compute nodes. This provides a way forward for seamless transition from physical clusters to virtual clusters, or for on-demand capacity bursting by adding virtual compute nodes to a physical cluster.

There are also many opportunities for performance tuning which by and large took a back seat during the implementation of this proof of concept, for example increasing the MTU.

ADVANTAGES OF OPENSTACK

Dave Holland, principal systems administrator at the Sanger Centre, adds:

"OpenStack brings flexibility—to developers, to users, and to system administrators. The ability to combine OpenStack and Lustre features in this way enables provisioning of familiar performant filesystems to users, with relatively low overhead compared to other solutions. This capability also offers an easier transition to 'cloudy' ways of working as we move forward to evaluating OpenStack-native solutions such as Manila."

Further Reading

An IBM research study on integrating Spectrum Scale (GPFS) within OpenStack environments: *http://www.redbooks.ibm.com/redpapers/pdfs/redp5331.pdf*

A 2015 presentation from ATOS on using Kerberos authentication in Lustre: *http://cdn.opensfs.org/wp-content/uploads/2015/04/Lustre-and-Kerberos_Buisson.pdf*

Glyn Bowden of HPE and Alex MacDonald from SNIA discuss OpenStack storage (including the Provisioned Filesystem Model using Lustre): *https://www.brighttalk.com/webcast/663/168821*

The High-Performance Big Data team at The Ohio State University: *http://hibd.cse.ohio-state.edu*

A useful talk from the 2016 Austin OpenStack Summit on Ceph design: https://www.openstack.org/videos/video/designing-for-high-performance-ceph-at-scale

The Ontario Institute for Cancer Research Collaboratory: http://www.cancercollaboratory.org

The billing and reporting apps developed at OICR are available at: https://github.com/CancerCollaboratory/billing

The Docker containers for the three components are published on Docker Hub: https://hub.docker.com/r/collaboratory/billing/

George Mihaiescu and Jared Baker presented Collaboratory at the 2017 Boston OpenStack summit: https://www.openstack.org/videos/boston-2017/operational-lessons-from-running-openstack-and-ceph-for-cancer-research-at-scale

Further details on the International Cancer Genome Consortium: http://icgc.org/

Dr. Tom Connor presented CLIMB at the 2016 Austin OpenStack Summit: https://www.openstack.org/videos/video/the-cloud-infrastructure-for-microbial-bioinformatics-breaking-biological-silos-using-openstack

OpenStack and HPC Workload Management

The approach taken for managing workloads is a major difference between HPC and conventional cloud use cases.

A typical approach to HPC workload management is likely to involve one or more head nodes of an HPC cluster for login, development, compilation and job submission services. Parallel workloads would be submitted from a head node to job batch queues of the workload manager, which control access to parallel partitions of compute nodes. Such partitions may equate to mappings of types of compute nodes and the specific resources (CPU, memory, storage and networking) that applications require. Each compute node runs a workload manager agent which configures resources, launches application processes, and monitors utilisation.

Pain Points in Typical HPC Workload Management

On a large, multi-user HPC system, the login node is a continual source of noisy neighbour problems. Inconsiderate users may, for example, consume system resources by performing giant compilations with wide task parallelism, open giant logfiles from their task executions, or run recursive finds across the filesystem to look for forgotten files.

An HPC system must often support a diverse mix of workloads. Different workloads may have a wide range of dependencies. With increasing diversity comes an increasing test matrix, which increases the toil involved in making any changes. How can an administrator be sure of the effects of any change to the software packages installed? What must be done to support a new version of an ISV application? What are the side effects of updating the version of a dependency? What if a security update leads to a dependency conflict? As the flexibility of an HPC software environment grows, so too does the complexity maintaining it.

In an environment where data is sensitive, local scratch space and parallel filesystems for HPC workloads can often have default access permissions with an undesirable level of openness. Data security can be problematic in a shared HPC resource in which the tenants are not trusted.

The Case for Workload Management on OpenStack Infrastructure

The flexibility of OpenStack can ease a number of pain points of HPC cluster administration:

- With software-defined OpenStack infrastructure, a new compute node or head node is created through software processes—not a trip to the data centre. Through intelligent, orchestrated automated provisioning, the administrative burden of managing changes to resource configuration can be eliminated. And from a user's perspective, a self-service process for resizing their resource allocation is much more responsive and devolves control to the user.

- Through OpenStack, it becomes a simple process to automatically provision and manage any number of login nodes and compute nodes. The multi-tenancy access control of cloud infrastructure ensures that compute resources allocated to a project are only seen and accessible to members of that project. OpenStack does not pretend to change the behaviour of noisy neighbours, but it helps to remove the strangers from a neighbourhood.

- OpenStack's design ethos is the embracing (not replacing) of data centre diversity. Supporting a diverse mix of HPC workloads is not materially different from supporting the breadth of cloud-native application platforms. One of the most significant advances of cloud computing has been in the effective management of software images. Once a user project has dedicated workload management resources allocated to it, the software environment of those compute resources can be tailored to the specific needs of that project without infringing on any conflicting requirements of other users.

- The cloud multi-tenancy implemented by OpenStack enforces segregation so that tenants are only visible to one another through the interfaces that they choose to expose. The isolation of tenants applies to all forms of resources—compute, networking and storage. The fine-grained control over what is shared (and what is not shared) results in greater data security than a conventional multi-user HPC system.

All of this can be done using typical HPC infrastructure and conventional cloud management techniques, but to do so would demand using industry best practices as a baseline, and require the continual attention of a number of competent system administrators to keep it running smoothly, securely, and to the satisfaction of the users.

Organisations working on the convergence of HPC and cloud often refer to this subject as Cluster-as-a-Service (CaaS). How can a cloud resource be equipped with the interfaces familiar to users of batch-queued conventional HPC resources?

Delivering an HPC Platform upon OpenStack Infrastructure

HPC usually entails a platform, not an infrastructure. How is OpenStack orchestrated to provision an HPC cluster and workload manager?

Addressing this market are proprietary products and open-source projects. The tools available in the OpenStack ecosystem also ensure that a home-grown cluster orchestration solution is readily attainable. An example of each approach is included here.

Broadly, the cluster deployment workflow would follow these steps:

1. The creation of the HPC cluster can be instigated through the command line. In some projects, a custom panel for managing clusters is added to Horizon, the OpenStack web dashboard.

2. Resources for the cluster must be allocated from the OpenStack infrastructure. Compute node instances, networks, ports, routers, images and volumes must all be assigned to the new cluster.

3. One or more head nodes must be deployed to manage the cluster node instances, provide access for end users, and workload management. The head node may boot a customised image (or volume snapshot) with the HPC cluster management software installed. Alternatively, it may boot a stock cloud image and install the required software packages as a secondary phase of deployment.

4. Once the head nodes are deployed with base OS and HPC cluster management packages, an amount of site-specific and deployment-specific configuration must be applied. This can be applied through instance metadata or a configuration management language such as Ansible or Puppet. A Heat-orchestrated deployment can use a combination of instance metadata and a configuration management language (usually Puppet but more recently Ansible provides such capability).

5. A number of cluster node instances must be deployed. The process of node deployment can follow different paths. Typically, the cluster nodes would be deployed in the same manner as the head nodes by booting from OpenStack images or volumes, and applying post-deployment configuration.

6. The head nodes and cluster nodes will share one or more networks, and the cluster nodes will register with the HPC workload management service deployed on the head nodes.

OPEN PLATFORMS FOR CLUSTER-AS-A-SERVICE

The simplest implementation is arguably ElastiCluster, developed and released as GPL open source by a research computing services group at the University of Zurich. ElastiCluster supports OpenStack, Google Compute Engine and Amazon EC2 as back-end cloud infrastructure, and can deploy (among others) clusters offering Slurm, Grid Engine, Hadoop, Spark and Ceph.

ElastiCluster is somewhat simplistic and its capabilities are comparatively limited. For example, it doesn't currently support Keystone v3 authentication—a requirement for deployments where a private cloud is divided into a number of administrative domains. A cluster is defined using an INI-format configuration template. When creating a Slurm cluster, virtual cluster compute nodes and a single head node are provisioned as VMs from the OpenStack infrastructure. The compute nodes are interconnected using a named OpenStack virtual network. All post-deployment configuration is carried out using Ansible playbooks. The head node is a Slurm controller, login node and NFS file server for /home mounting onto the compute nodes.

BRIGHT COMPUTING CLUSTER ON DEMAND

Bright Computing has developed its proprietary products for HPC cluster management and adapted them for installation, configuration and administration of OpenStack private clouds. The solution is capable of partitioning a system into a mix of bare metal HPC compute and OpenStack private cloud.

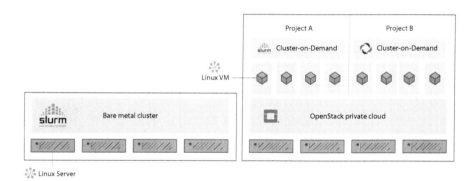

Bright Computing also provides an OpenStack distribution with a Bright-themed OpenStack web interface and an additional panel for management of Cluster on Demand deployments.

Cluster on Demand for OpenStack uses Heat for orchestrating the allocation and provisioning of virtualised cluster resources. When a virtual cluster is created, the Nova flavors (virtualised hardware templates) for head node and cluster compute node are specified. OpenStack networking details are also provided. Bright OpenStack is capable of deploying OpenStack with SR-IOV support, and Cluster on Demand is capable of booting cluster compute nodes with SR-IOV networking.

Cluster on Demand deployment begins with pre-built generic head node images. Those can then be quickly instantiated (via optional copy-on-write semantics) and automatically customised to user's requirements. Bright's deployment solution differs slightly from other approaches by using Bright Cluster Manager on the virtualised head node to deploy the virtual cluster nodes as though they were bare metal. This approach neatly nests the usage model of Bright Cluster Manager within a virtualised environment, preserving the familiar workflow of bare metal deployment. At the same time, the approach can utilise the efficiencies of the cloud infrastructure by pre-provisioning the virtual cluster's nodes from a pre-created generic node image (via optional copy-on-write semantics). Later, during the initial bootup process of the node, the node's filesystem is quickly customised by Bright to reflect the node's unique identity (e.g., some config files are updated, services get started). A virtualised cluster of "typical" size can be deployed on-demand from scratch in several minutes, at which point it is ready to accept HPC jobs.

Cluster on Demand focuses on delivering the flexibility advantages of self-service cluster provisioning, but can also deliver HPC performance. To minimise the network overheads of virtualisation, a cluster can be provisioned from compute nodes using SR-IOV. Using OpenStack Ironic, a cluster can be deployed with bare metal compute compute nodes, eliminating virtualisation altogether. Cluster on Demand supports the creation of heterogeneous clusters, composed with a combination of these approaches, to create a resource that is tailored to the workload's requirements and optimised for the infrastructure available.

A distinctive feature of Bright OpenStack is the ability to deploy virtualised HPC compute nodes next to physical ones, and run HPC workloads in an environment spanning a mixture of physical and virtual contexts. This capability to integrate conventional HPC infrastructure with on-demand cloud infrastructure enables new flexibility. Queues associated with each resource would enable the separate treatment of high-priority and low-priority HPC jobs. Long-running jobs can be virtualised, enabling their live migration to preserve them from the disruption of data centre changes.

Building on this capability, Bright supports cloud bursting when conventional HPC infrastructure is overwhelmed by demand—both locally to OpenStack resources and across the internet to other public clouds. When more HPC resources are needed during spikes in demand, they can be dynamically provisioned using cloud infrastructure. OpenStack's vision of a shared pool of compute resources, managed dynamically to serve the infrastructure needs across an organisation, now also includes the peak demand bursts of HPC. Wider usage helps reduce the TCO of an OpenStack cloud, and allows better utilisation of the hardware resources within it.

For greater convenience, Bright provides packaged configurations for a wide range of platforms that build upon Cluster on Demand: workload managers, big data services (Spark, Hadoop), deep learning tools, or even virtualised OpenStack clouds (OpenStack on OpenStack).

EXTENDING SLURM AND OPENSTACK TO ORCHESTRATE MVAPICH2-VIRT CONFIGURATION

The NOWLAB group at Ohio State University has developed a virtualised variant of their MPI library, MVAPICH2-Virt. MVAPICH2-Virt is described in greater detail in the *OpenStack and HPC Network Fabrics* section.

NOWLAB has also developed plugins for Slurm, called Slurm-V, to extend Slurm with virtualization-oriented capabilities such as submitting jobs to dynamically created VMs with isolated SR-IOV and inter-VM shared memory (IVSHMEM) resources. Through MVAPICH2-Virt runtime, the workload is able to take advantage of the configured SR-IOV and IVSHMEM resources efficiently. The NOWLAB model is slightly different from the approach taken in CaaS, in that a MVAPICH2-Virt-based workload launches into a group of VMs provisioned specifically for that workload.

> *"The model we chose to create VMs for the lifetime of each job seems a clear way of managing virtualized resources for HPC workloads. This approach can avoid having long-lived VMs on compute nodes, which makes the HPC resources always in the virtualised state. Through the Slurm-V model, both bare-metal and VM-based jobs can be launched on the same set of compute nodes since the VMs are provisioned and configured dynamically only when the jobs need virtualised environments."* —Prof. DK Panda and Dr. Xiaoyi Lu of NOWLAB

The IVSHMEM component runs as a software device driver in the host kernel. Every parallel workload has a separate instance of the IVSHMEM device for communication between co-resident VMs. The IVSHMEM device is mapped into the workload VMs as a paravirtualised device. The NOWLAB team has developed extensions to Nova to add the connection of the IVSHMEM device on VM creation, and recover the resources again on VM deletion.

Users can also hotplug/unplug the IVSHMEM device to/from specified running virtual machines. The NOWLAB team provides a tool with MVAPICH2-Virt (details can be found in MVAPICH2-Virt user guide *http://mvapich.cse.ohio-state.edu/userguide/virt/#_support_for_integration_with_openstack_for_vms*) to hotplug an IVSHMEM device to a virtual machine and unplug an IVSHMEM device from a virtual machine.

The Slurm-V extensions have been developed to work with KVM directly. However, the NOWLAB group have extended their project to enable Slurm-V to make OpenStack API calls to orchestrate the creation of workload VMs. In this model of usage, Slurm-V uses OpenStack to allocate VM instances, isolate networks and attach SR-IOV and IVSHMEM devices to workload VMs. OpenStack has already provided scalable and efficient mechanisms for creation, deployment, and reclamation of VMs on a large number of physical nodes.

Slurm-V is likely to be one of many sources competing for OpenStack-managed resources. If other cloud users consume all resources, leaving Slurm-V unable to launch sufficient workload VMs, then the new submitted jobs will be queued in Slurm to wait for available resources. As soon as one job completes and the corresponding resources are reclaimed, Slurm will find another job in the queue to execute based on the configured scheduling policy and resource requirements of jobs.

COMBINING THE STRENGTHS OF CLOUD WITH HPC WORKLOAD MANAGEMENT

At Los Alamos National Lab (LANL), there is a desire to increase the flexibility of the user environment of their HPC clusters.

To simplify their workload, administrators want every software image to be the same, everywhere. LANL systems standardise on a custom Linux distribution, based on Red Hat 7 and tailored for their demanding requirements. Sustaining the evolution of that system to keep it current with upstream development whilst maintaining local code branches is an ongoing challenge.

The users demand ever-increasing flexibility, but have requirements that are sometimes contradictory. Some users have applications with complex package dependencies that are out of date or not installed in the LANL distribution. Some modern build systems assume internet access at build time, which is not available on LANL HPC clusters. Conversely, some production applications are built from a code base that is decades old, and has dependencies on very old versions of libraries. Not all software updates are backwards compatible.

Tim Randles, a senior Linux administrator and OpenStack architect at the Lab, uses OpenStack and containers to provide solutions. Woodchuck is the LANL third-generation system aimed at accommodating these conflicting needs. The 192-node system has a physical configuration optimised for data-intensive analytics: a large amount of RAM per CPU core, local disk for scratch space for platforms—such as HDFS and Ceph—and 10G Ethernet with VXLAN, and SDN capabilities for virtualised networking.

Reid Priedhorsky at LANL has developed an unprivileged containerised runtime environment, dubbed "Charliecloud", upon which users can run applications packaged using Docker tools. This enables users to develop and build their packages on their (comparatively open) laptops or workstations, pulling in the software dependencies they require.

One issue arising from this development cycle is that in a security-conscious network such as LANL, the process of transferring application container images to the HPC cluster involves copying large amounts of data through several hops. This process was soon found to have drawbacks:

- It quickly became time consuming and frustrating.
- It could not be incorporated into continuous integration frameworks.
- The application container images were being stored for long periods of time on Lustre-backed scratch space, which has a short data retention policy, was occasionally unreliable, and not backed up.

Randles's solution was to use OpenStack Glance as a portal between the user's development environment on their workstation and the HPC cluster. Compared with the previous approach, the Glance API was accessible from both the user's workstation and the HPC cluster management environment. The images stored in Glance were backed up, and OpenStack's user model provided greater flexibility than traditional UNIX* users and groups, enabling fine-grained control over the sharing of application images.

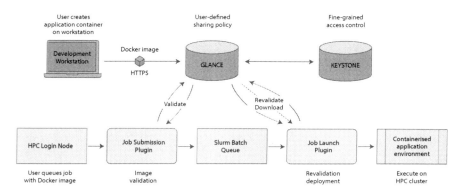

Tim developed Slurm plugins to interact with Glance for validating the image and the user's right to access it. When the job was scheduled for execution, user and image were both revalidated and the application image downloaded and deployed, ready for launch in the Charliecloud environment.

Giving users the ability to deploy containerized runtime frameworks introduced an additional challenge. Many of these frameworks utilize a client-server model for distributed task execution, yet lack any mechanisms for securing access. Therefore, it is possible for a user to start up a framework with daemons that have full access to their data but no access controls, exposing that user's data to anyone else able to submit workloads to the framework. The solution to this problem was to isolate a user's nodes on the cluster network.

Building on his success with Glance, Tim developed additional Slurm plugins to interact with Neutron to give Slurm the ability to create layer 2 VLANs to isolate user jobs from one another. These plugins also allowed LANL to dynamically grant and revoke compute node access to network-attached resources, such as databases and specialised filesystems, by manipulating compute node VLAN membership.

LANL is in the process of making the Glance and Neutron plugins available as open source contributions to Slurm's codebase.

HPC AND CLOUD CONVERGE AT THE UNIVERSITY OF MELBOURNE

Research compute clusters are typically designed according to the demands of a small group of influential researchers representing an ideal use case. Once built, however, the distribution of use cases can change as a broader group of researchers come onboard. These new uses cases may not match the expected ideal, and in some cases, conflict. If job queues and computation times stretch out, it can drive the proliferation of isolated department-level clusters, which are more expensive to maintain, lack scale, and are all too often orphaned when the responsible researcher moves on.

Introducing Spartan

In 2016, the University of Melbourne launched a new cluster called Spartan. It takes an empirical approach, driven by the job profiles observed in its predecessor, Edward, in the prior year. In particular, single-core and low-memory jobs dominate: 76 percent were single core, and 97 percent used <4GB of memory. High-core count, task-parallel jobs were often delayed due to competition with these single-core jobs, leading to research funds being directed towards department-level resources. National peak facilities were often rejected as an option due to their long queue times and restrictive usage requirements.

Spartan takes advantage of the availability of an existing and very large research cloud (NeCTAR) to allow additional computation capacity, and the provisioning of common login and management infrastructure. This is combined with a small but more powerful partition of tightly coupled bare metal compute nodes, and specialist high-memory and GPU partitions.

This hybrid arrangement offers the following advantages:

- Users with data parallel jobs have access to the much larger research cloud and can soak up the spare cycles available with cloud bursting, reducing their job wait time.
- Users with task parallel jobs have access to optimised bare metal HPC, supported by high-speed networking and storage.
- The larger task parallel jobs remain segregated from less resource-intensive data parallel jobs, reducing contention.
- Job demands can be continually monitored, and the cloud and bare metal partitions selectively expanded as and when the need arises.
- Departments and research groups can co-invest in Spartan. If they need more processing time or a certain type of hardware, they can attach it directly

to Spartan and have priority access. This avoids the added overheads of administering their own system, including the software environment, login and management nodes.

- Management nodes can be readily migrated to new hardware, allowing upgrades or hardware replacements without bringing the entire cluster down.
- Spartan can continue beyond the life of its original hardware, as different partitions are resized or replaced, a common management and usage platform remains.

Spartan does not have extraordinary hardware or software, and it's peak performance does not exceed that of other HPC systems. Instead, it seeks to segregate compute loads into partitions with different performance characteristics according to their demands. This will result in shorter queues, better utilisation, cost-effectiveness, and, above all, faster time to results for our research community.

Job and Resource Management

Previous HPC systems at the university utilised Moab Workload Manager for job scheduling and Terascale Open-source Resource and QUEue Manager (TORQUE) as a resource manager. The Spartan team adopted the Slurm Workload Manager for the following reasons:

- Existing community of users at nearby Victorian Life Sciences Compute Initiative (VLSCI) facility.
- Similar syntax to the PBS scripts used on Edward, simplifying user transition.
- Highly configurable through add-on modules.
- Support for cloud bursting, for example, to the OpenStack-based NeCTAR research cloud in Spartan's case or to Amazon Elastic Computing Cloud (EC2).

Account Management

Integration with a central staff and student active directory was initially considered, but ultimately rejected due to the verbose usernames required (i.e., email addresses). The Spartan team reverted to using an LDAP-based system as had been the case with previous clusters, and a custom user management application.

Application Environment

EasyBuild was used as a build and installation framework, with the LMod environmental modules system selected to manage application loading by users. These tools tightly integrate, binding the specific toolchains and compilation environment to the applications loaded by users. EasyBuild abstraction in its scripts sometimes required additional administrative overhead, and not all software had a pre-canned script ready for modification, necessitating them to be built from scratch.

Training

Training has been a particular focus for the implementation of Spartan. Previous HPC training for researchers was limited, with only 38 researcher days of training conducted in the 2012-2014 period. The Spartan team now engage in weekly training, rotating across the following sessions:

- Introductory, targeting researchers with little or no HPC or Linux experience.
- Transition, targeting existing Edward users who need to port their jobs to Spartan.
- Shell scripting.
- Parallel programming.

The team collaborates closely with researchers to drive this curriculum, serving a range of experience levels, research disciplines, and software applications.

Original Predictions

Bernard Meade, Spartan project sponsor, adds:

> *"The future configuration of Spartan will be driven by how it is actually used. We continue to monitor what applications are run, how long they take, and what resources they require. While Spartan has considerable elasticity on the cloud side, we're also able to incrementally invest in added bare metal and specialist nodes (high memory, GPU) as the need arises. Given the diversity in HPC, job characteristics will only grow; we believe this agile approach is the best means to serve the research community."*

This prediction has been verified. Shortly after official launch, there was a significant demand from one project for approximately double the memory per core from our original deployment. It was a relatively simple matter, relative to underlying hardware, to make this change. Since then the demands for single-node higher memory systems has increased, and the University has been able to make incremental purchases that suits these needs.

GPGPU Expansion

Recently, the University has been awarded a LIEF grant for significant GPGPU deployment. Management decided to deploy these GPGPUs as part of Spartan in their own partition. This will increase the processing power of the system as a whole from being a relatively small and experimental system, to one that is among the top five hundred public systems in the world.

Cloud Infrastructure Does Not (yet) Provide All the Answers

OPENSTACK CONTROL PLANE RESPONSIVENESS AND JOB STARTUP

Implementations of HPC workload management that create new VMs for worker nodes for every job in the batch queue can have consequential impact on the overall utilisation of the system if the jobs in the queue are comparatively short-lived:

- Job startup time can be substantially increased. A fast boot for a VM could be of the order of 20 seconds. Similarly, job cleanup time can add more overhead while the VM is destroyed and its resources harvested.
- A high churn of VM creation and deletion can add considerable load to the OpenStack control plane.

The CaaS pattern of virtualised workload managers does not typically create VMs for every workload. However, the OpenStack control plane can still have an impact on job startup time, for example if the application image must be retrieved and distributed, or a virtual tenant network must be created. Empirical tests have measured the time to create a virtual tenant network to grow linearly with the number of ports in the network, which could have an impact on the startup time for large parallel workloads.

WORKLOAD MANAGERS OPTIMISE PLACEMENT DECISIONS

A sophisticated workload manager can use awareness of physical network topology to optimise application performance through placing the workload on physical nodes with close network proximity.

On a private cloud system such as OpenStack, the management of the physical network is delegated to a network management platform. OpenStack avoids physical network knowledge and focuses on defining the intended state, leaving physical network management platforms to apply architecture-specific configuration.

In a CaaS use case, there are two scheduling operations where topology-aware placement could be usefully applied:

- When the virtual cluster compute node instances are created, their placement is determined by the OpenStack Nova scheduler.
- When a queued job in the workload manager is being allocated to virtual cluster compute nodes.

Through use of Availability Zones, OpenStack Nova can be configured to perform a simple form of topology-aware workload placement, but without any hierarchical grouping of nodes. Nova's scheduler filter API provides a mechanism which could be used for implementing topology-aware placement in a more intelligent fashion.

OPENSTACK'S FLEXIBILITY IS STRETCHED BY THE ECONOMICS OF UTILISATION

With its decoupled execution model, batch queue job submission is an ideal use case for off-peak compute resources. The AWS spot market auctions time on idle cores for opportunistic usage at up to a 90 percent discount from the on-demand price.

There is no direct equivalent to the AWS spot market in OpenStack. More generally, management of pricing and billing is considered outside of OpenStack's scope. OpenStack does not currently have the capabilities required for supporting opportunistic spot usage.

However, work is underway to implement the software capabilities necessary for supporting preemptible spot instances, and it is hoped that OpenStack will support this use case in due course. At that point, CaaS deployments could grow or shrink in response to the availability of underutilised compute resources on an OpenStack private cloud.

THE DIFFICULTY OF FUTURE RESOURCE COMMITMENTS

HPC facilities possess a greater degree of oversight and coordination, enabling users to request exclusive advance reservations of large sections of an HPC system to perform occasional large-scale workloads.

In private cloud, there is no direct mainstream equivalent to this. However, the Blazar project aims to extend OpenStack compute with support for resource reservations. Blazar works by changing the management of resource allocation for a segregated block of nodes. Within the partition of nodes allocated to Blazar, resources can only be managed through advance reservations.

A significant drawback of Blazar is that it does not support the intermingling of reservations with on-demand usage. Without the ability to gracefully preempt running instances, Blazar can only support advance reservations by segregating a number of nodes exclusively for that mode of usage.

Summary

OpenStack delivers new capabilities to flexibly manage compute clusters as on-demand resources. The ability to define a compute cluster and workload manager through code, data and configuration plays to OpenStack's strengths.

With the exception of some niche high-end requirements, OpenStack can be configured to deliver CaaS with minimal performance overhead compared with a conventional bare metal HPC resource.

Further Reading

The ElastiCluster project from the University of Zurich is open source. Online documentation is available here: *https://elasticluster.readthedocs.io/en/latest/index.html*

Bright Computing presented their proprietary Bright OpenStack and CaaS products at the 2016 Austin OpenStack Summit: *https://www.openstack.org/videos/video/bright-computing-high-performance-computing-hpc-and-big-data-on-demand-with-cluster-as-a-service-caas*

The NOWLAB's publication on Slurm-V: Extending Slurm for Building Efficient HPC Cloud with SR-IOV and IVShmem: *http://link.springer.com/chapter/10.1007/978-3-319-43659-3_26*

Tim Randles from Los Alamos presented his work on integrating Slurm with Glance on the HPC/Research speaker track at the 2016 Austin OpenStack Summit in April: *https://www.openstack.org/videos/video/glance-and-slurm-user-defined-image-management-on-hpc-clusters*. His work integrating Slurm with Neutron was presented at the 2016 Barcelona OpenStack Summit in October: *https://www.openstack.org/videos/barcelona-2016/neutron-and-slurm-software-defined-networking-for-hpc-clusters*

The Spartan OpenStack/HPC system at the University of Melbourne: *http://newsroom.melbourne.edu/news/new-age-computing-launched-university-melbourne* and *http://insidehpc.com/2016/07/spartan-hpc-service/*

Topology-aware placement in Slurm is described here: *http://slurm.schedmd.com/topology.html*

Some research describing a method of adding topology-aware placement to the OpenStack Nova scheduler: *http://charm.cs.illinois.edu/newPapers/13-01/paper.pdf*

HPC resource management at CERN and some OpenStack pain points as of April 2016 are described in detail here: *http://openstack-in-production.blogspot.co.uk/2016/04/resource-management-at-cern.html*

OpenStack Pre-emptible Instances Extension (OPIE) from Indigo Datacloud is available here: *https://github.com/indigo-dc/opie*

OpenStack and HPC Infrastructure Management

In this section, we discuss the emerging OpenStack use case for management of HPC infrastructure. We introduce Ironic, the OpenStack bare metal service, and describe some of the differences, advantages and limitations of managing HPC infrastructure as a bare metal OpenStack cloud.

Compared with OpenStack, established approaches to HPC infrastructure management are very different. Conventional solutions offer much higher scale, and much lower management plane overhead. However, they are also inflexible, difficult to use, and slow to evolve.

Through differences in the approach taken by cloud infrastructure management, OpenStack brings new flexibility to HPC infrastructure management:

- OpenStack's integrated support for multi-tenancy infrastructure introduces segregation between users and projects that require isolation.
- The cloud model enables the infrastructure deployed for different projects to use entirely different software stacks.
- The software-defined orchestration of deployments is assumed. This approach, sometimes referred to as "infrastructure as code", ensures that infrastructure is deployed and configured according to a prescriptive formula, often maintained under source control in the same manner as source code.
- The range of platforms supported by Ironic is highly diverse. Just about any hardware can and has been used in this context.
- The collaborative open development model of OpenStack ensures that community support is quick and easy to obtain.

The "infrastructure as code" concept is also gaining traction among some HPC infrastructure management platforms that are adopting proven tools and techniques from the cloud infrastructure ecosystem.

Deploying HPC Infrastructure at Scale

HPC infrastructure deployment is not the same as cloud deployment. A cloud assumes large numbers of users, each administering a small number of instances compared to the overall size of the system. In a multi-tenant environment, each user may use different software images. Without coordination between the tenants, it would be unusual for a cloud to deploy more than a few instances at any one time. The software architecture of the cloud deployment process is designed around this assumption.

Conversely, HPC infrastructure deployment has markedly different properties:

- A single user (the cluster administrator). HPC infrastructure is a managed service, not user-administered.
- A single software image. All user applications will run in a single common environment.
- Large proportions of the HPC cluster will be deployed simultaneously.
- Many HPC infrastructures use diskless compute nodes that network-boot a common software image.

In the terminology of the cloud world, a typical HPC infrastructure deployment might even be considered a "black swan event". Cloud deployment strategies do not exploit the simplifying assumptions that deployments are usually across many nodes using the same image and for the same user. Consequently, OpenStack Ironic deployments tend to scale to the low thousands of compute nodes with current software releases and best-practice configurations. Network booting a common image is a capability that only recently has become possible in OpenStack and has yet to become an established practice.

Bare Metal Management Using OpenStack Ironic

Using Ironic, bare metal compute nodes are automatically provisioned at a user's request. Once the compute allocation is released, the bare metal hardware is automatically decommissioned and ready for its next use.

Ironic requires no presence on the compute node instances that it manages. The software-defined infrastructure configuration that would typically be applied in the hypervisor environment must instead be applied in the hardware objects that interface with the bare metal compute node: local disks, network ports, etc.

SUPPORT FOR A WIDE RANGE OF HARDWARE

A wide range of hardware is supported, from full-featured BMCs on enterprise server equipment down to devices whose power can only be controlled through an SNMP-enabled data centre power strip.

An inventory of compute nodes is registered with Ironic and stored in the Ironic node database. Ironic records configuration details and current hardware state, including:

- Physical properties of the compute node, including CPU count, RAM size, and disk capacity.
- The MAC address of the network interface used for provisioning instance software images.
- The hardware drivers used to control and interact with the compute node.
- Details needed by those drivers to address this specific compute node (for example, BMC IP address and login credentials).
- The current power state and provisioning state of the compute node, including whether it is in active service.

INVENTORY GROOMING THROUGH HARDWARE INSPECTION

A node is initially registered with a minimal set of identifying credentials—sufficient to power it on and boot a ramdisk. Ironic generates a detailed hardware profile of every compute node through a process called Hardware Inspection.

Hardware Inspection uses this minimal bootstrap configuration provided during node registration. During the inspection phase, a custom ramdisk is booted, which probes the hardware configuration and gathers data. The data is posted back to Ironic to update the node inventory. Large amounts of additional hardware profile data are also made available for later analysis.

The inspection process can optionally run benchmarks to identify performance anomalies across a group of nodes. Anomalies in the hardware inspection dataset of a group of nodes can be analysed using a tool called Cardiff. Performance anomalies, once identified, can often be traced to configuration anomalies. This process helps to isolate and eliminate potential issues before a new system enters production.

BARE METAL AND NETWORK ISOLATION

The ability for Ironic to support multi-tenant network isolation is a new capability, first released in OpenStack's Newton release cycle. This capability requires some mapping of the network switch ports connected to each compute node. The mapping of an Ironic network port to its link partner switch port is maintained with identifiers for switch and switch port. These are stored as attributes in the Ironic network port object. Currently, the generation of the network mapping is not automated by Ironic.

Multi-tenant networking is implemented through a configuring state in the attached switch port. The state could be the access port VLAN ID for a VLAN network, or VTEP state for a VXLAN network. Currently, only a subset of Neutron drivers are able to perform the physical switch port state manipulations needed by Ironic. Switches with VXLAN VTEP support and controllable through the OVSDB protocol are likely to be supported.

Ironic maintains two private networks of its own: Networks dedicated to node provisioning and cleaning networks are defined in Neutron as provider networks. When a node is deployed, its network port is placed into the provisioning network. Upon successful deployment, the node is connected to the virtual tenant network for active service. Finally, when the node is destroyed, it is placed on the cleaning network. Maintaining distinct networks for each role enhances security, and the logical separation of traffic enables different QoS attributes to be assigned for each network.

Current Limitations of Ironic Multi-tenant Networking

In HPC hardware configurations, compute nodes are attached to multiple networks. Separate networks dedicated to management and high-speed data communication are typical.

Current versions of Ironic do not have adequate support for attaching nodes to multiple physical networks. Multiple physical interfaces can be defined for a node, and a node can be attached to multiple Neutron networks. However, it is not possible to attach specific physical interfaces to specific networks.

Consequently, with current capabilities only a single network interface should be managed by Ironic. Other physical networks would be managed outside of OpenStack's purview, but will not benefit from OpenStack's multi-tenant network capabilities as a result. Furthermore, Ironic only supports a single network per physical port: all network switch ports for Ironic nodes are access ports. Trunk ports are not yet supported, although this feature is in the development backlog.

REMOTE CONSOLE MANAGEMENT

Many server management products include support for remote consoles, both serial and video. Ironic includes drivers for serial consoles, built upon support in the underlying hardware.

Recently developed capabilities within Ironic have seen bare metal consoles integrated with the OpenStack Nova framework for managing virtual consoles. Ironic's node kernel boot parameters are extended with a serial console port, which is then redirected by the BMC to serial-over-LAN. Server consoles can be presented in the Horizon web interface in the same manner as virtualised server consoles.

Currently this capability is only supported for IPMI-based server management.

SECURITY AND INTEGRITY

When bare metal compute is sold as an openly accessible service, privileged access is granted to a bare metal system. There is substantial scope for a malicious user to embed malware payloads in the BIOS and device firmware of the system.

Ironic counters this threat in several ways:

- **Node Cleaning:** The Ironic node state machine includes states where hardware state is reset and consistency checks can be run to detect attempted malware injection. Ironic's default hardware manager does not support these hardware-specific checks. However, custom hardware drivers can be developed to include BIOS configuration settings and firmware integrity tests.
- **Network Isolation:** Through using separate networks for node provisioning, active tenant service, and node cleaning, the opportunities for a compromised system to probe and infect other systems across the network are greatly reduced.
- **Trusted Boot:** Use of a Trusted Platform Module (TPM) and chain of trust built upon it is necessary. These processes are used to secure public cloud deployments of Ironic-administered bare metal compute today.

None of these capabilities is enabled by default. Hardening the Ironic security model requires expertise and some amount of effort.

PROVISIONING AT SCALE

The cloud model use case makes different assumptions to HPC. A cloud is expected to support a large number of individual users. At any time, each user is assumed to make comparatively small changes to their compute resource usage. The HPC infrastructure use case is dramatically different. HPC infrastructure typically runs a single software image across the entire compute partition, and is likely to be deployed jointly in one operation.

Ironic's current deployment models do not scale as well as the models used by conventional HPC infrastructure management platforms. xCAT uses a hierarchy of subordinate service nodes to fan out an iSCSI-based image deployment. Rocks cluster toolkit uses BitTorrent to distribute RPM packages to all nodes. In the Rocks model, each deployment target is a torrent peer. The capacity of the deployment infrastructure grows alongside the number of targets being deployed.

However, the technologies for content distribution and caching that are widely adopted by the cloud can be incorporated to address this issue. Caching proxy servers can be used to speed up deployment at scale.

With appropriate configuration choices, Ironic can scale to handle deployment to multiple thousands of servers.

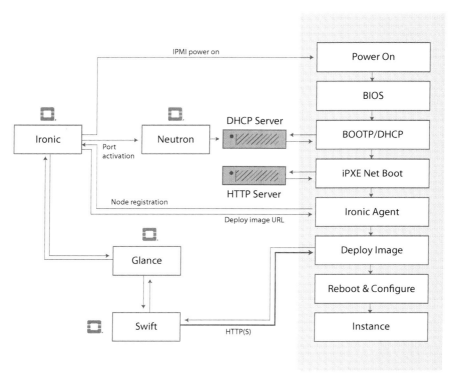

An overview of Ironic's node deployment process when using the Ironic Python Agent ramdisk and Swift URLs for image retrieval. This strategy demonstrates good scalability, but the deploy disk image cannot be bigger than the RAM available on the node.

Building Upon Ironic to Convert Infrastructure into HPC Platforms

The strengths of cloud infrastructure tooling become apparent once Ironic has completed deployment. From this point, a set of unconfigured compute nodes must converge into the HPC compute platform required to meet the user's needs. A rich ecosystem of flexible tools is available to perform this purpose.

See the *OpenStack and HPC Workload Management* section for further details of some of the available approaches.

CHAMELEON: AN EXPERIMENTAL TESTBED FOR COMPUTER SCIENCE

Chameleon is an infrastructure project implementing an experimental testbed for Computer Science led by University of Chicago, with Texas Advanced Computing Center (TACC), Renaissance Computing Institute (RENCI), and Northwestern University as partners. The Chameleon project is funded by the United States National Science Foundation (NSF). The project started in 2014 and is now in its second phase until 2020.

The current system comprises ~600 nodes split between sites at TACC in Austin and University of Chicago. The sites are interconnected with a 100G network. The compute nodes are divided into 12 racks, referred to as "standard cloud units", comprising 42 compute nodes, four storage nodes with 16 2TB hard drives each, and 10G Ethernet connecting all nodes with an SDN-enabled top-of-rack switch. Each SCU has 40G Ethernet uplinks into the Chameleon core network fabric. On this largely homogenous framework were grafted heterogenous elements allowing for different types of experimentation. One SCU has Mellanox ConnectX-3 InfiniBand interconnects. Two computer nodes were set up as storage hierarchy nodes with 512GB of memory, two Intel P3700 NVMe of 2TB each, four Intel S3610 SSDs of 1.6TB each, and four 15K SAS HDDs of 600GB each. Two additional nodes are equipped with NVIDIA Tesla K80 accelerators, two nodes with NVIDIA Tesla M40 accelerators, and sixteen nodes with two NVIDIA Tesla P100 accelerators each.

Heterogeneous cloud units for experimentation with alternate processors have been incorporated, including FPGAs, and microservers based on Intel Xeon, Intel Atom and ARM 64 microservers.

In the near term, additional nodes with GPU accelerators will be added to Chameleon, as these types of nodes have proved very popular.

Chameleon's public launch was at the end of July 2015; since then it has supported close to 300 research projects into computer science and cloud compute.

The system is designed to be deeply reconfigurable and adaptive, to produce a wide range of flexible configurations for computer science research. Chameleon uses the OpenStack Blazar project to manage advance reservation of compute resources for research projects. Since the OpenStack Summit in Barcelona in 2016, the Blazar project was revived by several institutions, including University of Chicago, NTT, and NEC. Blazar is now being considered for inclusion as an official OpenStack project.

Chameleon deploys OpenStack packages from RDO, orchestrated using OpenStack Puppet modules. Chameleon's management services currently run CentOS 7 and

OpenStack Liberty. Through Ironic, a large proportion of the compute nodes are provided to researchers as bare metal (a few SCUs are dedicated to virtualised compute instances using KVM). Chameleon's Ironic configuration uses the popular driver pairing of PXE-driven iSCSI deployment and IPMItool power management.

Ironic's capabilities have expanded dramatically in the two years since Chameleon first went into production, and many of the new capabilities are being integrated into this project.

The Chameleon project's is currently upgrading its OpenStack installation from the Liberty version (released in October 2015) to the Ocata version (released in February 2017), with the goal of providing the following features to users:

- *Network isolation* – placing different research projects onto different VLANs to minimise their interference with one another. Chameleon hosts projects researching radically different forms of networking, which must be segregated.
- *Bare metal consoles* – enabling researchers to interact with their allocated compute nodes at the bare metal level, with integration in the Horizon web interface.
- *BIOS parameter management* – enabling researchers to (safely) change BIOS parameters, and then to restore default parameters at the end of an experiment.

The Chameleon project is also investigating new capabilities included in Pike (the most recent OpenStack release at the time of writing; released in August 2017), such as Ironic-Cinder integration.

Pierre Riteau, DevOps lead for the Chameleon projects, sees Chameleon as an exciting use case for Ironic, which is currently developing many of these features.

"With the Ironic project, OpenStack provides a modern bare metal provisioning system benefiting from an active upstream community, with each new release bringing additional capabilities. Leveraging Ironic and the rest of the OpenStack ecosystem, we were able to launch Chameleon in a very short time.

Our decision to build Chameleon using OpenStack really paid off. From the Juno release we started with, Ironic has been steadily adding new features and gaining in maturity, to the point that most of our wish list has now been implemented. We are now in the process of integrating many of these new features in Chameleon, allowing us to make our testbed more powerful— thanks to the contributions of the OpenStack community.

We see the future of OpenStack in this area as providing a fully-featured system capable of efficiently managing data centre resources, from provisioning operating systems to rolling out firmware upgrades and identifying performance anomalies."

BRIDGES: A NEXT-GENERATION HPC RESOURCE FOR DATA ANALYTICS

Bridges is a supercomputer at the Pittsburgh Supercomputer Center funded by the NSF. It is designed as a uniquely flexible HPC resource, intended to support both traditional and non-traditional workflows. The name implies the system's aim, to "bridge the research community with HPC and Big Data."

Bridges supports a diverse range of use cases, including graph analytics, machine learning, and genomics. As a flexible resource, Bridges supports traditional Slurm-based batch workloads, Docker containers and interactive web-based workflows.

Bridges has 846 compute nodes, 16 of which have dual K80 GPU accelerators and 32 of which have dual P100 GPU accelerators from NVIDIA. There are also 46 high-memory nodes, including four with 12TB of RAM each. The entire system is interconnected with an Omnipath high-performance 100G network fabric.

Bridges is deployed using community-supported free software. The OpenStack control plane is CentOS 7 and Red Hat Distribution of OpenStack RDO (a freely available packaging of OpenStack for Red Hat systems). OpenStack deployment configuration is based on the PackStack project. Bridges was deployed using OpenStack Liberty and has been recently upgraded to OpenStack Ocata.

Most of the nodes are deployed in a bare metal configuration using Ironic. Puppet is used to select the software role of a compute node at boot time, avoiding the need to re-image. For example, a configuration for Message Passing Interface (MPI), Hadoop or virtualisation could be selected according to workload requirements. Additionally, Slurm reservations can be used to dynamically carve off a section of Bridges for virtualisation use.

OmniPath networking is delivered using the OpenFabrics Enterprise Distribution (OFED) driver stack. Compute nodes use IP over Open Platform for Architecture (OPA) for general connectivity. HPC apps use RDMA verbs to take full advantage of OmniPath capabilities.

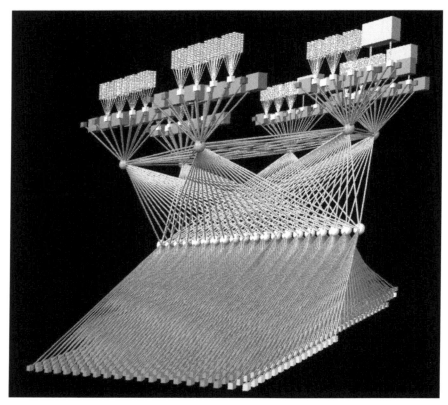

Visualisation of the Bridges OmniPath network topology. Eight hundred general purpose compute nodes and GPU nodes are arrayed along the bottom of the topology. Special purpose compute nodes, storage and control plane nodes are arrayed across the top of the topology. Forty-two compute nodes connect to each OmniPath ToR switch (in yellow), creating a "compute island", with 7:1 oversubscription into the upper stages of the network.

Bridges Exposes Issues at Scale

The Bridges system is a very large deployment for Ironic. While there are no exact numbers, Ironic has been quoted to scale to thousands of nodes.

Previously, coherency issues between Nova Scheduler and Ironic could arise if too many nodes were deployed simultaneously. These issues were addressed and not encountered after updating to an Ocata installation.

Within Ironic, the periodic polling of driver power states is serialised by each Ironic conductor. BMCs can be very slow to respond, and this can lead to the time taken to poll all power states in series to grow quite large. On Bridges, the polling takes approximately eight minutes to complete. This can also lead to apparent inconsistencies of state between Nova and Ironic, and the admin team work around this issue by enforcing "settling time" between deleting a node and reprovisioning it. With the move to Ocata, the Bridges framework is scheduled be upgraded to multiple conductors to address the polling and synchronization issues. Additionally, a multiple conductor setup will drastically decrease the deployment times for large scale system upgrades.

Benefiting from OpenStack and Contributing Back

The team at PSC have found benefits from using OpenStack for HPC system management:

- The ability to manage system image creation using OpenStack tools such as diskimage-builder.
- Ironic's automation of both PXE node and local node booting.
- The prescriptive repeatable deployment process developed by the team using Ironic and Puppet.

Robert Budden, senior cluster systems developer at PSC, has many future plans for OpenStack and Bridges:

- Using other OpenStack services such as Zun (containers) for both user and infrastructure needs.
- Developing Ironic support for network deploy and boot over OmniPath.
- Diskless boot of extremely large memory nodes using the Ironic-Cinder integration.
- Deployment of a dynamically scaling OpenStack Ironic control plane.
- Increased convergence between bare metal and virtualised OpenStack deployments.
- Ironic and Neutron integration for network tenant isolation.
- Developing native Omnipath plugins for Neutron.
- Developing Manila support for Lustre and Slash2 filesystems.

Budden adds:

> "One of the great things is that as OpenStack improves, Bridges can improve. As these new projects come online, we can incorporate those features and the Bridges architecture can grow with the community.
>
> A big thing for me is to contribute back. I'm a developer by nature, I want to fix some of the bugs and scaling issues that I've seen and push these back to the OpenStack community."

PROTOTYPING FUTURE WORKLOADS FOR RADIO ASTRONOMY

It is an exciting time to be a radio astronomer. The Square Kilometre Array (SKA) is an ambitious multinational project that promises to push the boundaries of radio telescope design further than ever before. When it enters service in the middle of the next decade, the SKA aims to probe deeper into space, and further back in time. This powerful instrument is being designed to unlock the secrets of phenomena such as pulsars, gravity, and the cosmic dawn.

The SKA project has two principal centres: a low-frequency telescope in the deserts of Western Australia and a mid-frequency telescope in the deserts of South Africa. In each desert region, the dishes and antennas connect with a Central Signal Processor (CSP) or correlator. Fibre links couple these installations with a dedicated supercomputer, the Science Data Processor (SDP), in a nearby city. The combined processing power of the SDP is estimated to be of order 250 PetaFLOPS—roughly the combined power of the twelve most powerful computers in the June 2017 TOP500 list.

To deliver on its ambitions, the SKA faces new challenges at every level. After initial processing, the sustained data rates between the telescope CSP and the SDP will be on the order of 1 Terabyte per second. Depending on the particular observations, up to six hours of this data must be stored and readily available to enable rapid analysis of the current sky. A near-realtime data analysis is also required to enable the telescope to detect and respond quickly to sky events. Six months of processed data must be held to enable further analysis. A federated network of regional centres is required to store, share and support access to this archive. All of this must be done on a scale previously unseen for a project of this kind.

None of this is easy, but through careful selection of the right frameworks, the SKA software architecture will become possible. The ALaSKA (à la SKA) project is a system designed to inform the design process.

Bare Metal OpenStack: the Crucible for Architectural Experiments

ALaSKA is an OpenStack private cloud designed and developed to embrace data centre diversity.

The system has 36 heterogeneous compute nodes, comprising a number of general-purpose nodes, plus nodes equipped with high-memory or multiple GPUs. Most nodes are equipped with multiple SSDs, but a number of alternative storage architectures are present, including nodes equipped with either 24 SSDs or high-performance NVMe devices. Underpinning the infrastructure's persistent storage is an ARM64-powered Ceph cluster. The system also has a number of network fabrics, including EDR (100Gbit) InfiniBand, together with 100G, 25G and 10G Ethernet.

This system is a performance prototype platform. Diverse software frameworks can be deployed, explored and evaluated on a system that embodies the current performance envelope. The researchers using the system are interested in programming with a range of software frameworks, without becoming entangled in the complexity of infrastructure and framework deployment.

OpenStack delivers on these requirements. ALaSKA uses Ironic to manage the compute nodes as bare metal, but remains a flexible infrastructure and offers simple deployment of a range of software frameworks using Ansible, Heat, Magnum and Sahara. For agility, the OpenStack control plane is containerised using Kolla, and deployed using Bifrost and Kayobe. In addition to controlling and configuring the servers, Kayobe also configures network switches, enabling the entire infrastructure to be version-controlled as source code.

ALaSKA was initially deployed using the OpenStack Ocata release, and has recently been upgraded to the Pike release in order to take advantage of recent developments in Ironic for managing bare metal infrastructure.

Some of the the future plans for ALaSKA include:

- Greater support for deeply-reconfigurable polymorphism: the ability to reconfigure the BIOS and RAID parameters of a compute node, depending on whether it has been deployed to run Slurm or Spark, for example.
- Integrating fine-grained performance monitoring, to provide data to users that can assist with performance analysis.
- During early operation the system quickly became fully utilised, and a method of managing resource reservations (such as Blazar) will likely be needed for effective ongoing management.

Most of the Benefits of Software-Defined Infrastructure

In the space of HPC infrastructure management, OpenStack's attraction is centred on the prospect of having all the benefits of software-defined infrastructure while paying none of the performance overhead.

To date, there is no single solution that can provide this. However, a compromising trade-off can be struck in various ways:

- Fully virtualised infrastructure provides all capabilities of cloud with much of the performance overhead of cloud.
- Virtualised infrastructure using techniques such as SR-IOV and PCI pass-through dramatically improves performance for network and I/O intensive workloads, but imposes some constraints on the flexibility of software-defined infrastructure.

- Bare metal infrastructure management using Ironic incurs no performance overhead, but has further restrictions on flexibility.

Each of these strategies is continually improving. Fully virtualised infrastructure using OpenStack private cloud provides control over performance-sensitive parameters like resource over-commitment and hypervisor tuning. It is anticipated that infrastructure using hardware device pass-through optimisations will soon be capable of supporting cloud capabilities, such as live migration. Ironic's bare metal infrastructure management is continually developing new ways of presenting physical compute resources as though they were virtual.

OpenStack has already arrived in the HPC infrastructure management ecosystem. Projects using Ironic for HPC infrastructure management have already demonstrated success. As it matures, its proposition of software-defined infrastructure without the overhead will become increasingly compelling.

A RAPIDLY DEVELOPING PROJECT

While it is rapidly becoming popular, Ironic is a relatively young project within OpenStack. Some areas are still being actively developed. For sites seeking to deploy Ironic-administered compute hardware, some limitations remain. Ironic has a rapid pace of progress, and new capabilities are released with every OpenStack release cycle.

HPC infrastructure management using OpenStack Ironic has been demonstrated at over 800 nodes, while Ironic is claimed to scale to managing thousands of nodes. However, new problems become apparent at scale. Currently, large deployments using Ironic should plan for an investment in the skill set of the administration team and active participation within the Ironic developer community.

Further Reading

A clear and helpful introduction into the workings of Ironic in greater detail: *http://docs.openstack.org/developer/ironic/deploy/user-guide.html*

Deployment of Ironic as a standalone tool: *http://docs.openstack.org/developer/bifrost/readme.html*

Kate Keahey from University of Chicago presented an architecture show-and-tell about Chameleon at the 2016 Austin OpenStack Summit: *https://www.openstack.org/videos/video/chameleon-an-experimental-testbed-for-computer-science-as-application-of-cloud-computing-1*

Chameleon Cloud's home page is at: *https://www.chameleoncloud.org*

Robert Budden presented an architecture show-and-tell on Bridges at the 2016 Austin OpenStack Summit: *https://www.openstack.org/videos/video/deploying-openstack-for-the-national-science-foundations-newest-supercomputers*

Further information on Bridges is available at its home page at PSC: *http://www.psc.edu/index.php/bridges*

Further information on the SKA Telescope: *http://skatelescope.org*

OpenStack and Research Cloud Federation

Science today is no longer exclusively produced in single research labs or within national boundaries. Modern scientific challenges call for collaborations with researchers from different institutions across different countries and access to computing power with flexible usage to analyse vast amounts of data. Cloud federations enable effective collaboration for researchers as they allow compute resources managed by one institution to be seamlessly accessible to collaborators from other institutions and vice versa.

Federated identity is one of the core parts of any federated infrastructure. With federated identity, authentication is securely delegated to the user's home institution credential provider instead of having to manage independent local accounts at each provider available to a scientific collaboration.

This section introduces the basic federated identity management concepts and showcases existing research cloud federations enabled with federated identity and OpenStack.

Federated Identity Management

Federated identity allows users of one domain to securely access systems of another domain with their origin domain credentials and without the need for redundant user administration. With federation identity:

- Local OpenStack admins do not need to provision and manage individual user accounts from collaborations, but instead rely on the existing users and the authentication mechanism available at their home institutions.
- Users are provided with single sign-on, so the same set of credentials give access to resources across several OpenStack deployments.

In federated identity, there is a trust relationship between a service provider (SP), in our case an OpenStack cloud, and an identity provider (IdP), a service that is able to authenticate and identify the users. Whenever a user wants to access a resource on the service provider, the SP initiates an authentication flow as follows:

1. The SP redirects the user to the IdP, requesting authentication.
2. The user is authenticated at the IdP. The authentication method depends on each IdP and this step may be skipped if the user was already logged in.

3. After successful authentication, the IdP redirects the user back to the SP with a set of attributes (assertions or claims) with further user information.

4. The SP authorizes the user depending on her attributes coming from the IdP and provides access to the resource.

Federated identity is enabled by open standards and/or openly published specifications, like SAML by OASIS or OpenID Connect (OIDC) by the OpenID Foundation. SAML is an XML-based protocol that uses security tokens with assertions about the end user between the IdP and SP. SAML federations are frequent in the research community and is the basis of EduGAIN. OpenID Connect leverages OAuth 2.0 authorisation framework and uses a RESTful HTTP API with JSON as a data format.

Keystone—OpenStack Identity Service: first introduced support for federated identity in the Icehouse release (2014.1) with the OS-FEDERATION extension of the Identity API v3.0, that enables IdP management, protocols (e.g. SAML or OIDC), and mappings, which are a set of rules evaluated to authorise a user depending on the attributes provided by the IdPs. Keystone's approach to federation is independent of the federated identity protocol used, and relies on external Apache plugins to provide the actual support for the different technologies.

Federated Identity at CERN

The CERN IT department provides compute resources to physicists throughout the world who work on CERN-approved experiments. In 2013, CERN's OpenStack cloud became a production service which has now grown to around 280,000 cores across two data centres.

Through the CERN/Rackspace collaboration within the CERN openlab, and community participation from multiple organisations, the OpenStack federated identity functionality was included as standard in the upstream open source project.

This functionality was demonstrated at the OpenStack summit in Hong Kong in 2013 (*https://www.openstack.org/videos/hong-kong-2013/hybrid-openstack-clouds-cern-research-project-aims-to-solve-federation-for-the-real-world*) accessing two OpenStack clouds using federated identity and has been significantly developed further in the four years since.

Most recently, CERN has been working within an EU Horizon 2020 project Indigo Datacloud (*https://www.indigo-datacloud.eu/*) to provide resources for testing the scientific platform-as-a-service. With 26 partners, establishing access to these resources rapidly at the start of the project was of significant interest. Using the GEANT EduGAIN identity provider, we federated access to CERN cloud resources.

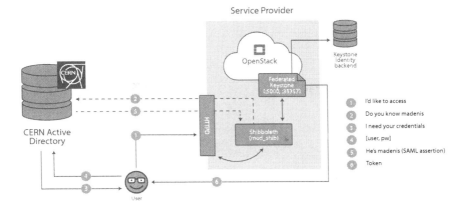

Every federated user is mapped to a group. The group has access to the project as members. Both the project and the group are owned by a CERN user that acts as the contact point for the federation link. The group is assigned to the user according to the attributes of their identity, using a combination of SAML assertions and the Keystone mappings.

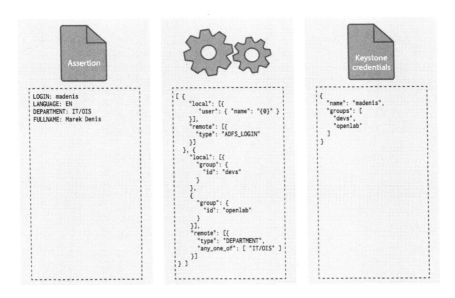

In order to simplify the setup, the CERN IdP (CERN Single Sign-on) acts as a proxy for the eduGAIN federation (only 1 IdP is registered in Keystone), and establishes a list of trusted attributes upon which we base our SAML assertion rules. This avoids the negotiation between IdPs on the attribute agreement and the exponential growth on the rules (1 or more per IdP).

As all our active accounts are handled by an identity management solution outside the OpenStack setup, these group mappings are also stored in it. Then the information is accessible to us via an LDAP interface to the Active Directory domain.

A local MySQL database is configured as the default backend since the federated domain is needed in addition to the Active Directory domain. The assertion rules and shadow accounts created by the federated users are stored in the local database and Active Directory is not modified.

In deploying this functionality in collaboration with the CERN security team, we identified two additional functionalities which were then enabled on the CERN cloud.

First, there is a need for traceability. This allows the administrators of the cloud service to be able to trace back to the original identity of the end user. Enabling Cloud Auditing Data Federation (CADF) in OpenStack provided the logs to ensure that this could be performed.

Secondly, compute resources at CERN have to be owned by an authorised user who has signed the CERN computing rules. This is enforced when a virtual machine is configured with a CERN-managed IP address. The CERN network driver was enhanced so that each project has a sponsor who has signed the computing rules and ensures the other members of the project are aware of their responsibilities.

With these functionalities, cloud federation using OpenStack and eduGAIN were deployed successfully in production.

Nordic e-Infrastructure Collaboration (NeIC)

The goal of the NeIC-funded Glenna project is to share knowledge and set best practices on managing cloud services and to create a Nordic federated cloud service, driven by the need of the Nordic researchers. The sharing of IaaS and SaaS resources in the Nordic countries enables an easier form of collaboration. Instead of moving large amounts of data between countries, the analysis can be done where the data is and Nordic researchers can run their experiments and manage their data on any Nordic cloud.

IaaS resources currently provided to users are exclusively OpenStack-based. The collaboration also provides access to PaaS/SaaS resources across the Nordics which are built on frameworks such as NextCloud and Galaxy.

Each country provides different clouds of services:

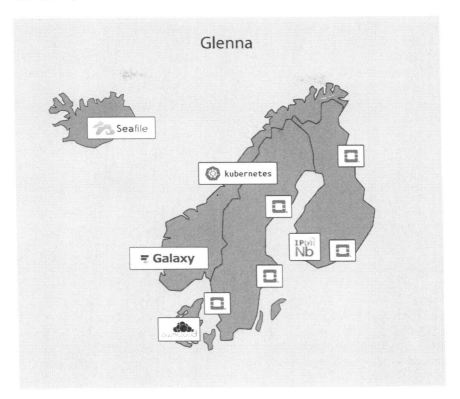

Single sign-on has, up until recently, relied on the Kalmar2 e-identity Union (*https://www.uninett.no/en/kalmar2*), a pan-Nordic infrastructure formed by the identity federation members HAKA in Finland, Feide in Norway, WAYF in Denmark and SWAMID in Sweden for authentication and authorisation of end users. Work has recently started on identifying and permitting access to users belonging to the international eduGAIN interfederation, which is expected to widen the scope of access options.

The Nordic cloud services primarily rely on the SAML 2.0 federated identity management protocol. The Nordic federations have had a national scope, and have been constructed around the national education and research networks. Entities participating in Kalmar2 were required to support SAML 2.0 as specified by OASIS. Furthermore, Kalmar2 specified a deployment profile defining how to configure the SAML 2.0 entities.

The Kalmar2 entities also had to follow the Interoperable SAML 2.0 Web Browser SSO Deployment Profile (*http://saml2int.org*), and in order to avoid conflicting IDs, each participating federation (country) was required to ensure that all participating entities have an SAML2.0 EntityID in a controlled namespace.

For true federated cross-organisational access, the key missing piece is commonly accepted group management services. The EduGAIN and Kalmar projects have gone a long way towards providing federated authentication, but the authorisation and resource management part is still lacking. In OpenStack, the resources are managed and quotas granted on the project level. Using only Kalmar and EduGAIN, there are no overreaching concepts that properly map to OpenStack projects. There are several third-party projects (e.g. Grouper (*https://www.internet2.edu/products-services/trust-identity/grouper/*), Géant EduTEAMS (*https://www.geant.org/Innovation/eduteams*), Cesnet Perun (*https://perun.cesnet.cz/web/*) and Indico IAM (*https://www.indigo-datacloud.eu/identity-and-access-management*)) that provide SAML-based user group management services, which would logically map to the project concept. None of these services are currently in wide use within the Nordics. One of the remaining challenges for a true federation that serves the research users' needs, is to have a shared service that can be used for this purpose. One aim for the Glenna project is to evaluate and integrate to one of these services to gain practical experience about how this works technically. Each participating service provider then faces the challenge of integrating this system with their internal processes.

Lack of policies associated with funding, security, and localization requirements remain key challenges, when it comes to the cross-border resource sharing. The loosely coupled approach of the Glenna project allows each federation to be fully autonomous in offering different services according to local policies and the available expertise and resources. The SNIC Science Cloud (SSC) in Sweden is one example where a national-scale cloud facility offers services according to the local needs, yet participates in the Glenna project to foster research needs across geographical boundaries. Currently, SSC also relies on SAML 2.0 and in the near future, will enable eduGAIN for the cross-border availability of resources within the Glenna project.

EGI Federated Cloud

EGI operates one of the largest e-infrastructures in the world to offer reliable ICT services, which provide uniform, cost-effective, user-oriented and collaborative access to computing and data storage resources in more than 30 countries. The EGI Federated Cloud, a subset of EGI's e-infrastructure, is a multi-national standards-based open cloud system that integrates institutional IaaS clouds into a computing platform for data and/or compute-driven applications and services.

EGI Federated Cloud is composed of independently operated resource providers that are integrated into the infrastructure by interacting with a set of components of the EGI Federation services for ensuring single-sign on; uniform interfaces (i.e., each provider supporting a given community can be accessed via the same/

harmonised APIs) and application portability (i.e., every provider supporting a community share the same VM images). There is no mandate to deploy a common cloud management stack, as long as these can be integrated with the EGI services and the IaaS interfaces provided to final users are agreed with the research collaborations. Most of the current members of the federation are operating OpenStack and offer both OpenStack and OCCI APIs to users. The Open Cloud Computing Interface (OCCI) is a RESTful protocol and API designed to facilitate interoperable access to, and query of, cloud-based resources across multiple resource providers and heterogeneous environments.

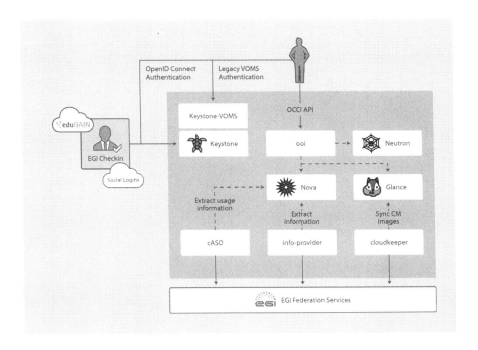

From the user's perspective, single sign-on is one of the core parts of any federation. Until recently, users accessed EGI access to resources employing personal X.509 certificates and the concept of Virtual Organisation (VO). Users create augmented proxy certificates signed with the VO attributes and with the user's personal certificate. The proxy and the different attributes determine at the resource provider level if the user is granted access to the resources. EGI provides the Keystone-VOMS (Virtual Organization Membership Service) module to enable this kind of authentication into OpenStack for Keystone V2. It relies on Keystone support for external authentication and is implemented as a WSGI filter of the server. The user authenticates against the HTTPD server with a VOMS proxy, and the server, after the proxy validation, includes the SSL information in the request environment. This information reaches the VOMS filter, that will try to authorise requests that include the attributes configured by the administrator. EGI also provides plugins for OpenStack clients so they can be easily used with proxies to interact with the federated resources.

EGI also provides Keystone V3 support for VOMS proxies relying exclusively on the OS-FEDERATION extension and the use of Apache modules, without the need to introduce any changes in the standard Keystone WSGI pipeline. Admins create mappings for authorisation with the attributes coming from the user's proxy (VO information) in the same way as with any other federated identity mechanism.

EGI CheckIn is the new EGI service for authentication and authorisation to access the EGI infrastructure without having to deal with X.509 certificates. Researchers from home organisations that participate in one of the eduGAIN federations are able to access the EGI services using the same credentials they are using at their home organisation by leveraging federated authentication mechanisms like SAML and OpenID Connect. The architecture of EGI CheckIn follows the guidelines from the AARC and AARC2 projects, where a central hub (EGI CheckIn) mediates between the federated IdPs and the EGI SPs. This decoupling reduces the complexity on the service providers, as they need to establish and maintain a technical and trust relationship with only a single entity, the EGI CheckIn, instead of managing many-to-many relationships. EGI CheckIn also handles and unifies the provision of attributes about users (e.g. VO membership) by contacting the relevant attribute authorities (e.g. VO management software) and including them in the assertions provided to the SPs. This allows providers to configure mappings for authorisation in a consistent way, independently of the VO or the IdP of the user.

OpenStack providers use OpenID Connect instead of SAML for authenticating EGI users. OpenID Connect has better support for non-browser-based access and opens the door for using not only Horizon but also command line tools or APIs directly, using the same credentials at every provider of the federation.

Integration with other services of EGI enhance the federation and is performed with a set of components that use the public OpenStack services APIs:

- cASO extracts usage information and publishes central accounting records where VOs can analyse and visualise usage of resources across the whole infrastructure.
- The info-provider extracts basic information about the providers, such as endpoints and supported VOs, so users can easily discover the available resources for their computations.
- CloudKeeper synchronises VM images from the EGI VM marketplace. Each VO can define a set of images that will be propagated automatically to every provider of the federation.
- ooi provides OCCI API support by translating user requests to the corresponding actions on Nova, Glance and Neutron OpenStack services.

Open Challenges in Federated Identity

Federated identity enables shared use of OpenStack clouds for research communities, but there are still a number of open challenges that need to be addressed.

As research collaborations evolve, new members join and others leave, but this information is not directly available to the local OpenStack deployments supporting the federation, as membership to collaboration is managed externally. How are active resources assigned to a user that is no longer in a collaboration dealt with? The System for Cross-domain Identity Management (SCIM) open standard was created to automate the exchange of user identity information between identity domains. SCIM could be used to automatically de-provision users at the local OpenStacks if IdPs and OpenStack implement support for them.

Once users are authenticated via the federated identity mechanism, they need to be authorised to access the OpenStack resources. This is performed by mapping the attributes of the user provider from the IdPs to the local OpenStack projects. These mappings are created and configured manually by administrators and need to be updated as the collaboration evolves, and also need to be consistent across the whole federation. Standards like the eXtensible Access Control Markup Language (XACML) could be used to define authorisation policies and distribute them.

Users may need to perform unattended authorised actions against a third party from an existing resource, e.g., accessing storage from a job running on a VM. In these situations, support for the delegation of credentials is essential so the existing resource can act on behalf of the user. The Token Exchange OAuth specification was designed to provide a protocol in support of these scenarios.

INDIGO-DataCloud Identity and Access Management (IAM) Service was developed to tackle these issues and provides support for SCIM, XACML and Token Exchange. The integration of SCIM and XACML in OpenStack would facilitate and improve the research federations.

Open Research Cloud

While federated identity enables shared use of OpenStack clouds for research communities, there are still obstacles that interfere in the ability of globally dispersed researchers to effectively collaborate. The Open Research Cloud (ORC), a collaboration of the international community supporting scientific research computing, was created to identify and remove those obstacles.

The ORC was formed and held its inaugural meeting in Boston in April 2017, co-located with the OpenStack Summit, with the ambition of bringing practitioners together to discuss those areas. The participants then signposted, in the form of a declaration, guidance to technology creators, operators, and vendors on how they may design or temper their cloud products and services to support federating capability of clouds used by researchers globally.

The initiative covers topics like the federation of data, identity, security, shared compute and storage. The ORC group has identified a range of issues that warrant further and more detailed exploration. These include: relative focus on policy vs. technical capabilities; dealing with the concerns about the loss of control and lock-in; data ownership/sovereignty and security; federation of identity management and protocols for access and authorised use (across multiple political jurisdictions); shared use, interoperability, mobility and provisioning; and addressing the combination of cultural and technical barriers to effective collaboration.

Once in final form, the declaration will be shared broadly with the global scientific research community and those institutions, organisations and commercial entities encouraged to adopt and support the conventions and guidance articulated by the community.

Further Reading

NeIC, Glenna Project *https://neic.no* and *https://neic.no/glenna2/*

SNIC Science Cloud *https://cloud.snic.se*

EGI AAI guide for OpenStack *https://wiki.egi.eu/wiki/AAI_guide_for_OpenStack*

Open Research Cloud *http://www.openresearchcloud.org/*

Summary

OpenStack cloud architecture and deployment offers software-defined, self-service infrastructure to the scientific community allowing researchers to maximise the time spent on research itself. Hundreds of scientific organisations are realising the benefits of OpenStack cloud computing for complex HPC workloads, including increased flexibility, adaptability, automation and management capabilities. Their willingness to contribute information about their architectures, operational criteria, and decisions made this paper possible.

This paper describes an in-depth view into the considerations for virtualisation, high-performance data and networking, and workload and infrastructure management. The intricacies are many, however, using OpenStack services and other open source and vendor technologies, the end users can be shielded from the complexity. The software is evolving to further meet the needs of multiple scientific use cases, and users—IT and scientists—are contributing code and experiences to accelerate the progress.

Each section includes *Further Reading*. The OpenStack community also provides planning and implementation documentation to delve deeper into OpenStack cloud software, architecture, and important topics such as networking and security. The active OpenStack scientific community and ecosystem is invaluable for their experience and advice, and is a great way to get involved. Visit openstack.org to get started or click on these resources for more information:

RESOURCE	OVERVIEW
OpenStack Scientific Working Group (*https://wiki.openstack.org/wiki/Scientific_working_group*).	Represents and advances the needs of research and high-performance computing atop OpenStack; provides a forum for cross-institutional collaboration. All are welcome to join.
OpenStack Operators mailing list (*http://lists.openstack.org/cgi-bin/mailman/listinfo/openstack-operators*; use [scientific-wg] in the subject.)	A forum for existing OpenStack cloud operators to exchange best practices around operating an OpenStack installation at scale.

RESOURCE	OVERVIEW
Internet Relay Chat (IRC) (https://wiki.openstack.org/wiki/IRC)	Scientific Working Group IRC meetings (online) are held in alternating time zones, in IRC channel #openstack-meeting: ■ Every two weeks on even weeks, 1100 UTC on Tuesday ■ Every two weeks on odd weeks, 0900 UTC on Wednesday
OpenStack Marketplace (http://www.openstack.org/marketplace/)	One-stop resource to the skilled global ecosystem for distributions, drivers, training, services and more.
OpenStack Architecture Design Guide (http://docs.openstack.org/arch-design/)	Guidelines for designing an OpenStack cloud architecture for common use cases. With examples.
OpenStack Networking Guide (https://docs.openstack.org/neutron/latest/admin/)	How to deploy and manage OpenStack Networking (Neutron).
OpenStack Security Guide (http://docs.openstack.org/security-guide/)	Best practices and conceptual information about securing an OpenStack cloud.
Complete OpenStack documentation (http://docs.openstack.org/)	Index to all documentation, for every role and step in planning and operating an OpenStack cloud.
Welcome to the community! (http://www.openstack.org/community/)	Join mailing lists and IRC chat channels, find jobs and events, access the source code and more.
User groups (https://groups.openstack.org/)	Find a user group near you, attend meetups and hackathons—or organize one!
OpenStack events (http://www.openstack.org/community/events/)	Global schedule of events including the popular OpenStack Summits and regional OpenStack Days.

Printed in Poland
by Amazon Fulfillment
Poland Sp. z o.o., Wrocław